# COMPUTERS
# IN ANALYTICAL
# CHEMISTRY

# PROGRESS IN
# ANALYTICAL CHEMISTRY
*Based upon the Eastern Analytical Symposia*

Series Editors:
Ivor L. Simmons
*M&T Chemicals, Inc., Rahway, N. J.*

and Paul Lublin
*General Telephone and Electronics Laboratories, New York, N.Y.*

Volume 1
H. van Olphen and W. Parrish
X-RAY AND ELECTRON METHODS OF ANALYSIS
*Selected papers from the 1966 Eastern Analytical Symposium*

Volume 2
E. M. Murt and W. G. Guldner
PHYSICAL MEASUREMENT AND ANALYSIS OF THIN FILMS
*Selected papers from the 1967 Eastern Analytical Symposium*

Volume 3
K. M. Earle and A. J. Tousimis
X-RAY AND ELECTRON PROBE ANALYSIS
IN BIOMEDICAL RESEARCH
*Selected papers from the 1967 Eastern Analytical Symposium*

Volume 4
C. H. Orr and J. A. Norris
COMPUTERS IN ANALYTICAL CHEMISTRY
*Selected papers from the 1968 Eastern Analytical Symposium*

PROGRESS IN ANALYTICAL CHEMISTRY
VOLUME 4

# COMPUTERS
# IN ANALYTICAL
# CHEMISTRY

Edited by Charles H. Orr
*Miami Valley Laboratory*
*The Procter & Gamble Co.*
*Cincinnati, Ohio*

and John A. Norris
*Baird-Atomic*
*Bedford, Massachusetts*

SPRINGER SCIENCE+BUSINESS MEDIA, LLC 1970

Library of Congress Catalog Card Number 78-89531

ISBN 978-1-4684-3317-3      ISBN 978-1-4684-3315-9 (eBook)
DOI 10.1007/978-1-4684-3315-9


© 1970 Springer Science+Business Media New York
Originally published by Plenum Pres, New York in 1970
Softcover reprint of the hardcover 1st edition 1970


United Kingdom edition published by Plenum Press, London
A Division of Plenum Publishing Company, Ltd.
Donington House, 30 Norfolk Street, London W.C. 2, England

# PREFACE

The analytical chemist is in the forefront of the race to use computers in laboratory work. The modern laboratory has a large number of instruments churning out information, and mechanized procedures for handling the huge amount of data are imperative. The marriage of instruments and computers is offered as a way of easing the burden on the scientist, as well as optimizing the performance of the analytical instruments. Computer systems can be applied to all the major analytical instrument procedures, and many of the leading instrument manufacturers are developing and producing systems for use in the laboratory, both for data acquisition and for control purposes.

It is, therefore, timely that the session on computers in analytical chemistry of the Eastern Analytical Symposium, held in November 1968, be published in this series, which has as its aim progress in analytical chemistry. The contents are wide-ranging and include applications to mass spectrometry, X-ray spectrography, nuclear magnetic resonance spectroscopy, gas chromatography, infrared spectrography, the use of dedicated computers, and the multiple user laboratory.

Thanks are due to the authors of the papers and to the session chairmen for their efforts in the production of this very worthwhile addition to the series.

Ivor L. Simmons
Paul Lublin

# CONTENTS

*Chapter I*
**An Approach to a Multiple User Laboratory** ............................... 1
    by Gary O. Walla
1. An Approach to a Multiple User Laboratory Automation System    1
2. Division of Responsibilities ............................................. 2
    2.1. Instrument Data Reduction and Analysis ......................... 2
    2.2. System Design and Implementation ............................... 3
    2.3. System Operation and Maintenance ............................... 4
3. Automation of a Specific Instrument ................................... 4
    3.1. Establishment of Capabilities and Objectives ................. 4
    3.2. Design of Hardware and Software ................................. 5
    3.3. Hardware and Software Maintenance ............................. 7
4. Summary ................................................................. 8
Appendix ................................................................... 8

*Chapter II*
**Dedicated Computer in the Laboratory** ............................... 11
    by Bradley Dewey III
1. General ................................................................. 11
2. Off-Line Computer ..................................................... 12
3. On-Line Computer ...................................................... 12
4. Functions of a Computer ............................................... 13
5. Advantages ............................................................. 15

*Chapter III*
**Real-Time High-Resolution Mass Spectrometry** ..................... 17
    by A. L. Burlingame, D. H. Smith, T. O. Merren, and R. W. Olsen
1. Introduction ........................................................... 17
2. Experimental ........................................................... 18
3. System Evaluation ..................................................... 19
    3.1. Calculation of $N$ ............................................... 19
    3.2. Mass Measurement Accuracy ....................................... 21
    3.3. Intensity Measurement Precision and Accuracy ................. 25

4. Applications to Organic Analysis ........................................... 27
Conclusions ............................................................................... 37
References ................................................................................. 37

*Chapter IV*

**A Computer Controlled X-ray Spectrograph** ............................. 39
    by Paul A. Weyler

1. Introduction ......................................................................... 39
2. Equipment ............................................................................ 40
3. Operating Variables .............................................................. 41
    3.1. X-ray Tube .................................................................... 41
    3.2. Sample ........................................................................... 42
    3.3 Goniometer Setting ($2\theta$) ...................................... 42
    3.4. Detector ......................................................................... 43
    3.5. Crystal ........................................................................... 43
    3.6. Collimator ..................................................................... 43
4. Computer Program ............................................................... 44
5. Sample Preparation—Geological Materials ........................... 46
6. Comments ............................................................................. 47
References ................................................................................. 48

*Chapter V*

**Computer Interface and Digital Sweep for an NMR Spectrometer**   49
    By Richard C. Hewitt

1. Introduction ......................................................................... 49
2. System Implementation ........................................................ 50
3. Detailed Circuit Description .................................................. 51
4. Conclusion ............................................................................ 62
References ................................................................................. 62

*Chapter VI*

**Application of the Infotronics CRS—110/50 Computer Integrator
Systems for On-Line GC Analyses** ............................................. 63
    by J. M. Cotton

1. Introduction ......................................................................... 63
2. Composite Integrator/Computer System ................................ 64
3. Modifications to the Basic System ......................................... 66
4. Programing the Integrator-Computer System ........................ 69
5. System Flexibility — Conclusions ........................................... 73

*Chapter VII*

**On-Line Operation of a PE 621 Infrared Spectrophotometer—
IBM/1800 Computer System** ................................................  75
    by T. Chuang, G. Misko, I. G. Dalla Lana, and D. G. Fisher
1. Introduction ..............................................................  75
2. System Components and Their Operation ..........................  78
    2.1. PE 621 Infrared Spectrophotometer .......................  79
    2.2. Encoder Readout-Computer Interface .....................  79
    2.3. IBM/1800 System ...........................................  80
    2.4. Operation of the System ....................................  83
3. Data Processing .........................................................  84
4. Research Applications ..................................................  88
    4.1. Measurement of Differential Spectra for Experiments on a
        Single Sample ................................................  88
    4.2. Infrared Monitoring of a Reaction System ...............  89
5. Comments ................................................................  90
Acknowledgment ...........................................................  92
References ...................................................................  92


*Chapter VIII*

**D ... Computer, Where's My Curve?** ..............................  93
    by W. R. Kennedy
References ...................................................................  103


**Index** .....................................................................  105

# I. AN APPROACH TO A MULTIPLE USER LABORATORY

Gary O. Walla

*The Procter & Gamble Company*
*Cincinnati, Ohio*

At the present time it is difficult to merge several scientifically oriented users and their specialized instruments together into one real-time computer using existing vendor supplied real-time operating systems. Considerable time and money can be wasted by trying to get all the scientific users trained in the detail of a real-time system. Likewise, it is not desirable to have the scientist work through a systems programmer every time he desires to make a slight (or major) change in a data analysis program. The approach discussed provides for a FORTRAN programing capability for the scientist which appears to him to be basically the same as any batch processing system, but yet is in the real-time environment processing data obtained directly from laboratory instruments.

## 1. AN APPROACH TO A MULTIPLE USER LABORATORY AUTOMATION SYSTEM

There are several approaches to installing a real-time data acquisition system. The approach taken will depend on factors such as data sampling rates, number and size of applications, and the cost of equipment and manpower. The most straightforward approach to automate a particular instrument is to use a small computer which is dedicated to only one instrument. This computer acquires the data from the instrument, processes it, and puts the output on a typewriter. It may also punch cards or paper tape for permanent record or for input to a larger computer.

When several instruments are to be automated, however, the economics and versatility of a centralized computer facility usually make it attractive to use a single medium-sized computer for automation. The card reader, line printer, and disk storage provided with most medium-sized process control computers reduces the total manpower effort required for programing and testing compared to the small computer. The approach discussed will be concerned with a centralized medium-sized computing facility where fast turnaround (less than 1 hr) batch FORTRAN processing is being done as well as simultaneous data acquisition from a number of laboratory instruments on a non-scheduled (demand) basis. The approach provides for a

1

high degree of sophistication and experimentation in the processing of data acquired from instruments, but is not designed for extensive development of new instruments. We recognize that in our environment the major productivity of a research chemist comes with the analysis of data acquired by an instrument, not in developing a new instrument.

In this approach, one or two people develop an expertise for the particular computer being used and they become responsible for keeping the system running and helping with new applications. The basic guideline for these "computer personnel" is to bring real-time intrument data to the "FORTRAN level." Thus, although freedom for the "scientist" to do anything he pleases with the computer is restricted, he can forget about the well-known details, including machine language programing, involved in the basic systems programing needed to acquire his data. He can then concentrate his skill and knowledge on analyzing the data using the computer through FORTRAN or other higher level languages or developing techniques to get new data.

In the Research Division of Procter & Gamble, we have been operating an IBM/1800 computer since late 1966 which simultaneously acquires data from several instruments. These instruments serve chemical research personnel who are adept at using computers and have developed all the programs necessary for analysis of their data. Since a "standard specification" for an interface between any analytical instrument and any digital computer does not exist, our scientists together with the computer lab staff (of which I am a member) have developed specifications in conjunction with equipment suppliers to connect specific instruments to the IBM/1800. In cases where vendors would not design instrument interfaces to the computer, the computer lab staff designed and installed interface hardware as well as the basic real-time data acquisition programs.

The operation of a medium-sized computer notably benefits from a division of responsibilities between the scientist, system analyst, and electronic designer because of the complexity and sophistication of the system. An approach to the division of responsibilities which has been successful in our operation is discussed below, and it complements the logical steps which are taken to automate an instrument.

## 2. DIVISION OF RESPONSIBILITIES

### 2.1. Instrument Data Reduction and Analysis

In real-time data acquisition, the data input rate is controlled by the instrument producing the data. This necessitates careful programing to insure that each incoming data point can be processed before the next point arrives. This is entirely different than a normal batch FORTRAN program where information is entered via cards or tape and is only read when the program requests input data. Data points being read from the card reader are time independent. In real-time acquisition the acquired information is not time independent, but can be made available on a time independent basis by storing the incoming data in essentially raw form in bulk storage (disk) until the "end of data" or "end of run" is sensed. Once the end of data signal is sensed, a logical stopping point has been reached and the information is no longer time dependent. Data can then be processed as any other data would be, in an ordinary batch computer environment, except that program execution will be scheduled by the "real-time" computing system instead of by an "EXECUTE" card in the input stream.

By using this technique, a scientist can program in FORTRAN with almost exactly the same facilities he uses in the batch environment except information comes from temporary storage (disk) instead of the card reader. The same memory and language size limitations apply to his special program as other normal batch processing programs. Also the same training and consulting facilities used for normal FORTRAN programing are available for use for real-time data reduction and analysis programs. Program development and testing can be done through the batch processing environment with confidence that real-time interactions will not affect the program. As a result, the scientist who knows FORTRAN will not have to wait for computer people to help him with his program and can proceed using his own time schedule.

## 2.2. System Design and Implementation

The instrument system design must be a joint effort between the scientist, system analyst, and electronic designer. The system analyst and electronic designer cannot and need not be expert in the application of every instrument they help automate, but they need a thorough understanding of the computer capabilities as well as a knowledge of electronic circuitry which is needed to interface with the computer.

The depth of understanding needed to automate a particular instrument is dependent on the type of commercially available equipment. Many instruments contain elaborate data gathering hardware and can be automated simply by transmitting the data, usually in digital form at this point, directly to the computer. Only transmission error detection procedures and signal level compatibility need to be considered for this class of instrument. Other instruments with basic analog output signals (such as GC and NMR) require investigation about sampling rates, band-width, and noise characteristics during the system design phase. With the scientist's basic understanding of the instrument and the system analyst's and electronic designer's understanding of the computer capability, the entire system can be defined usually with a minimum of outside consultation. After the overall system is designed, the system analyst and electronic designer work out the details and install the necessary hardware and software (programing) to acquire the data. This is done parallel and independently of the analysis programs being concurrently developed by the scientist. It is the analyst's responsibility to insure that this particular data acquisition system is compatible with and can co-exist simultaneously with other systems which might be running at the same time.

## 2.3. System Operation and Maintenance

The unsung hero of the computing system is the man who keeps the machine in a productive state once all these wonderful new capabilities have been installed. A computer is only useful when the program is in an operative, available condition. The ease with which a computer program can be changed enables it to be experimented with and developed to a very sophisticated level. There are many positive benefits to be gained by these changes. Therefore, the computing system must be capable of adapting to a situation of constant change. It cannot be stressed enough that an underlying assumption to the entire computing environment is that change should be expected and welcomed! An overall system which is designed to easily accept new or different programs, techniques, and expansion will have a much better chance of success in the research environment than a system designed for a fixed situation.

There are two key considerations in creating a flexible reliable system. Programs (such as those in the Appendix) which might be used for multiple instruments should have provisions for expansion. Thus, when the actual expansion occurs, it can be done with only a loss of a day of system operation to make changes as compared to several days if all programs have to be changed and checked out at the time of expansion. The other key consideration is to be sure adequate manpower is available to organize, update, and file the program source cards, program documentation and backup systems for the entire computing facility. The day to day dependency of laboratory personnel on the computing facility requires that a considerable effort be made to keep the system running even though changes are being made. This necessitates a considerable program maintenance effort.

## 3. AUTOMATION OF A SPECIFIC INSTRUMENT

### 3.1. Establishment of Capabilities and Objectives

The most important part of automating an instrument is the definition of the overall capabilities and objectives. Too often we are so anxious to get something working that we forget to plan for the overall system. This results in having to redesign parts of the system or having to live with a system which does not provide the capabilities desired.

A good place to start in defining capabilities is to look at existing instruments, either computer automated or not, and determine which capabilities are desirable and which are not. Undesirable limitations imposed by the instrument should be investigated to see if these are limitations imposed by physical laws or limitations imposed by economic considerations. Since the computer has available considerable logic at a relatively low cost, many things previously too expensive to be done in electronic hardware become economical using the computer. Defining capabilities initially, without concern for limitations, is important as experience has shown that many limitations which exist only for economic reasons are assumed by the scientist to be invariants.

Although most instruments being automated at this time are basically used in the same way as their hardware predecessors, the new future capabilities of the instrument should be kept in mind. If this instrument is primarily a research tool for a scientist instead of a production tool for routine analysis, then one of the objectives defined for automation might very well be the capability to easily add to or change data recording techniques or detection algorithms. A little imagination directed toward future capabilities can often provide information which will tip the balance one way or the other for some design decisions.

By the time all the capabilities and objectives for a specific instrument are defined, the data to be acquired and the computer control required over the instrument will be very well understood by everyone. The next step is to determine the best way to accomplish these objectives within the computing environment available.

## 3.2. Design of Hardware and Software

Today there is no such thing as a standard instrument–computer interface. Perhaps in the future there will be. Until that time, instrument-computer interfaces must be tailored to the individual computing environment. Computing environment implies more than just voltage level compatibility, it also includes acquisition (interrupt) response time, as well as instrument control response time and error checking. Typically, the faster the response time, the more expensive the application.

On our IBM/1800 system, response times of 10 sec or so are done through core swapping. Core swapping is transferring the current program to disk, loading the requested program from disk, executing the program, then reloading and resuming execution of the original program. These slow response programs can be written in FORTRAN with the program size only of secondary importance. Responses which must occur within a period shorter than 10 sec must have permanently resident programs in core at all times. Programs located permanently in core, requiring a dedicated part of memory, must be short; hence, they are programed in assembly language by a systems programer. These programs are not easily changed because the system must be taken out of operation for the change.

Response time is taken into consideration when designing an interface. Rapid data input usually requires very little, if any, program logic resulting in a small program (under 200 words) which resides permanently in memory. Control which requires fast response is also designed with very little program logic and as a result is small. For example, the logic to stop a drive motor consists of counting down a preset number of computer interrupts which are generated by the instrument–computer interface for each revolution (or 1000 revolutions, or seconds, or some other event). When the count reaches zero the motor is stopped within milliseconds. The logic needed to start the motor can be quite involved as it may consider which motor to drive, at what speed, and how far it must go. The interface can be designed so that computer response to start the system can be very slow (5–30 sec). This program can be written in FORTRAN without serious size limitations (8000 words) and can make very complex logical decisions which could involve data stored in files on the disk. When the program calculates how far this particular motor is to travel, it stores the calculated interrupt counts in the small resident program and starts the motor. The combination of small fast-response programs supporting the large slow-response programs has provided for all the capabilities we have desired to date.

A fast-response program is short and runs quickly. Still there is an upper limit on the speed of interrupt responses. One limitation is the computer manufacturer's electronic filtering on the interrupt input signal lines. The IBM/1800 limits a single interrupt to a rate of 100–200 interrupts/sec. This is adequate for most applications, but for those that are extremely fast, IBM will provide the electronic hardware which will operate at much higher rates. Unfortunately, however, although the electronic hardware will interrupt at high rates, the computer programs usually contain too many instructions to execute that quickly. Even with fast executing programs there is another difficulty. Sometimes it is imperative that the computer program currently executing not be interrupted, and thus the interrupts are inhibited (masked) for some period of time. The maximum inhibit time observed on our IBM/1800 (with a 2-$\mu$sec memory access time and IBM TSX executive program) is 1 msec. Considering all the filter delay time, possible interrupt mask time, and the time it takes the system program to get to the interrupt service routine, a reasonable estimate of the fastest interrupt rate was established at 500/sec (one interrupt every 2 msec).

Another aspect of system design is related to the inherent binary architecture of the computer. For many years instruments have been designed for data output to humans, not computers. If the system is designed using the computer as an integral part of the instrument and the primary communication link to the human, a lot of time, effort, and money may be saved. One example of this is the use of a shaft encoder to provide shaft angular position. A BCD (binary coded decimal) encoder has been used in the past because its output can be displayed directly in decimal to the human. A similar encoder using gray code, however, is less expensive, more reliable, and provides the information to the computer in a much more readable and compact form (base 2 instead of base 10). However, the human cannot easily and directly read the gray code, but by using the computer as the link between the encoder and the human (via a typewriter or other communication device) the required capabilities are obtained at a lower cost and with the possibility of higher sophistication.

### 3.3. Hardware and Software Maintenance

Once an instrument has been put into operation with the computer, it is important that it stay operating. Since these systems run for long periods without attention, a person's memory cannot be depended on to remember all the details of the system. It becomes quickly evident that written documentation is required if the system is expected to continue to operate effectively. The thoroughness required of documentation depends very much on how critical the instrument is to the overall operation. Will having the instrument off for an hour, day, or week dramatically hurt overall operation? The level of documentation is closely associated with the speed with which an instrument can be repaired when it fails.

Good electronic test equipment is essential to a good performing system. It is especially recommended that a good storage oscilloscope be available for testing equipment connected to the computer. Computer interface signals are usually slow and nonperiodic which makes them difficult to view with a conventional oscilloscope, yet they are too fast for meters or indicators. In one of the first instruments we automated, we had difficulty with the end of data interrupt. Sometimes it worked right, sometimes it didn't. The manufacturer, using a conventional oscilloscope, showed us the end of data interrupt as it left the interface. He was satisfied the problem was in the computer. It was only after we displayed both the data and end of data signals on a storage oscilloscope over a 2-sec sample of signal that we could demonstrate that the interface occasionally failed to transmit an end of data when the instrument was in a certain operating mode. With this information the manufacturer was able to find and correct logic error in his interface.

## 4. SUMMARY

There has purposely been no discussion of the software system which we developed to operate with IBM's TSX system to automate our instruments. The reason for this is simple; the system itself is not the key to making the operation work. It was the recognition and acceptance of the division of responsibilities by the participating people that made the system work. Our scientists who write the data reduction and analysis programs are current in their respective fields and are at the frontier in chemical research. The system analysts, such as myself, take the computer programs and equipment commercially available as a starting point and add the necessary programing needed to meet the requirements of our scientists. We do not do basic research in computing. The programing systems we develop are new to the scientist, but they are standard computer techniques being applied to a special situation. For purposes of illustration, I have included in the Appendix some basic subroutines which could be programed utilizing the facilities of IBM's TSX system. These are types of subroutines our programmers use for data acquisition.

I believe the success of a medium-sized computing system is dependent on how well the three areas of responsibilities are accepted and executed. Each area presents a unique challenge to the individuals concerned and to neglect any area will seriously degrade the system.


## APPENDIX

This appendix presents an example of how TSX system subroutines can be adapted to a multiple user laboratory automation system. The details of implementation will vary from system to system but can be easily developed for a given computer by the systems programmer. This example is concerned with the setting or resetting the digital output voltage levels which provide instrument control or signaling from the IBM/1800. The important characteristics of these subroutines are the ease with which they can be called from a FORTRAN program, the protection provided by the system for invalid data, and the ease with which the number of digital outputs can be changed with a major system change.

The hardware structure of the IBM/1800 with its 16-bit register configuration puts some limitations on the flexibility of the Digital Output (DO) facilities. Since all 16 bits are "set" or "rest" at one time, care must be taken when programs on two different interrupt levels share the 16 bits of the DO. Even if both programs are on the same interrupt level, the sharing of a digital output necessitates that each program knows what the other program is putting on the DO.

The problem of shared Digital Outputs as well as easy to use FORTRAN subroutines can be solved by specialized DO subroutines. First, general subroutines are provided which are linked together and serve as the medium which updates the DO table with the current bit configuration in each DO. Second, the subroutines program to a common interrupt level which provides all communication to the IBM DAOP subroutine. Any programmer using DO must use these subroutines and not call DAOP directly to maintain the integrity of the system. The use of these subroutines limits the use of the Digital Output to single-word operation, but for laboratory automation this is no detriment.

CALL DOBIT (NUM,NBIT,NX)        "Digital Output of a Bit"
                                Subroutine

A *single* DO bit is turned on or off with this subroutine. NUM is the DO address. The acceptable address limits are in the subroutine and the specified address is checked for validity every time the subroutine is called. NBIT is the bit number to be turned on or off. A value of one in NX will turn number NBIT bit on and a zero will turn the bit off.

CALL DOWRD (NUM,NWORD,NBITS)        "Digital Output of a
                                    Word" Subroutine

If more than one bit in a DO is to be turned on and/or off simultaneously then this subroutine is called. Again, NUM identifies the DO address. The argument NBITS is used to identify which bits in the DO are to be set (either on or off). The argument NBITS should have a "1" in each bit position which is to be used for output. NWORD has a "1" in each bit which is to be "on" and a "zero" in each bit that is to be "off". Only the bits which are identified in NBITS with a "1" will be looked at in NWORD.

EXAMPLES:
    CALL DOBIT (111,13,1)
        Turn on bit 13 of DO 111
    CALL DOBIT (109,0,0)
        Turn off bit 0 of DO 109
    NBITS = 7      (0000 0000 0000 0111)
                   (Binary bits are arranged by using a
                   FORTRAN DATA statement).
    NWORD = 2      (0000 0000 0000 0010)

    CALL DOWRD (111,NWORD,NBITS)
        Turn off bit 13 of DO 111
        Turn on bit 14 of DO 111
        Turn off bit 15 of DO 111

All other bits remain unchanged from previous condition.

Note: All bits in this case are changed simultaneously.

NBITS = 240      (0000 0000 1111 0000)
NWORD = 160   (0000 0000 1010 0000)

CALL DOWRD (105,NWORD,NBITS)
    Turn on bits 8 and 10 of DO 105
    Turn off bits 9 and 11 of DO 105

All other bits remain unchanged.

Since the digital address is an actual machine hardware address, electronics for digital outputs can be added to the 1800 system at any time without having to change any programs except the digital output subroutines and then it is only a change of the limits of valid addresses, a very minor change. As different bits of a 16-bit group are assigned to different devices there is no problem with interaction between devices because these subroutines take care of all interactions. Further, the FORTRAN programmer is independent of others and only needs his DO addresses and bits within the word.

# II. DEDICATED COMPUTER IN THE LABORATORY

Bradley Dewey III

*Digital Equipment Corporation*
*Maynard, Massachusetts*

The true small computer dedicated to a specific function in a laboratory has been responsible for a revolution in the methodology, quantity, economy, and quality of experiments performed. When connected to analytical instruments, the dedicated small computer becomes an element in a total analytical instrumentation system and is capable of acquiring and analyzing data, and controlling the instrument and experiment based on the data received.

Laboratory systems built around the dedicated small computer are already used in a variety of disciplines, such as X-ray diffraction, mass spectroscopy, gas chromatography, nuclear magnetic resonance spectrometry, and signal averaging. Prepackaged computer-and-software systems are available as "black box" devices for scientists who do not want to design their own systems.

When desired, the analytical and computational capability of the dedicated small computer-based system may be expanded by interfacing them to larger computers, which are then used as peripheral processors.

## 1. GENERAL

The small, general-purpose computer dedicated to a specific laboratory function has been responsible for a revolution in the methodology, quantity, economy, and quality of experiments performed. When interfaced to analytical instruments, the dedicated small computer becomes an element in a total analytical instrumentation system which is capable of acquiring and analyzing data as well as controlling the instrument and experiment based on the data received.

Until recently, scientists were restricted in the data they could handle by that which they could record by hand, or such data which was recorded automatically would have to be reduced by hand or programed for off-line processing by a large batch processing computer.

The advent of on-line computers, both dedicated and time shared, permits considerably more sophisticated instrumentation systems to be used that was heretofore possible. In an instrument such as the exhaust emission analyzer, the computer has actually become an integral part of the instrument.

Before going any further, we should define the different classes of computers. The basic classifications which I have already mentioned are on-line and off-line.

## 2. OFF-LINE COMPUTER

An off-line computer is one which accepts the experimental data from either magnetic tape, cards, punched paper tapes, or a typewriter. The criterion for an off-line computer is that it is not taking its data directly from the analytical instrument or process in "real-time."

## 3. ON-LINE COMPUTER

An on-line computer is one that is taking the data directly from the instruments or process and recording it, analyzing it, reducing it, and perhaps making decisions based on it. The on-line computer may be dedicated, or it may be a time sharing system operating multiple instruments or processes while at the same time performing other tasks such as batch processing.

The difference between these two types of computers (the on-line and the off-line) should be obvious. With the off-line computer you can only determine the past history of an experiment, whereas with the on-line computer you can often determine results in sufficient time to investigate new and unusual occurrences during the course of an experiment. When the on-line computer is dedicated to the individual instrument or process, turn around time between the end of the experiment and the time in which the results are available is vastly decreased. These results are often available in preliminary form even before the instrument or process has finished its cycle.

There are three basic relationships between on-line computers and analytical instruments: the built-in computer; the single purpose dedicated computer; and the time-shared computer. Let me explain each of these in order.

    1. As the name implies, the built-in computer is often buried inside the instrument or process. Such a self-contained system effectively does only one job. Sometimes it is only by having such a computer system that a job can be done at all. An example of this might be a system designed to analyze the exhausts from automotive engines. This instrument is built around one of our computers. Without the

computer it would be almost impossible to make a meaningful analytical relationship between the various spectrometers incorporated in the exhaust emission analyzer.

2. The second form of computer (the single-purpose dedicated computer) is one working with only one instrument or one type of instrument. The instrument might be a high resolution mass spectrometer, or multiple gas chromatographs, but in either case it is an instrument whose function is sufficiently complex that it may completely tie up the use of the computer. Such a computer does have the advantage of being able to be used for other functions when the instrument to which it is dedicated is not in operation. It may, for example, run gas chromatographs during the day and time average for an NMR spectrometer during the evening.

3. The third type of computer is one which is dedicated not to a single instrument or type instrument but to a group of different instruments. This is the time-shared system. In addition to operating with multiple instruments, a time-shared computer system should be capable of running batch processes, and debugging and running user written programs. Virtually by definition, this time-shared computer is a medium-to-large scale computer costing upwards of several hundred thousand dollars.

## 4. FUNCTIONS OF A COMPUTER

Before going any farther in analyzing the role of a dedicated computer, we should discuss the three functions that a computer performs for laboratory instrumentation. These functions are: control, data acquisition, and data analysis.

Computer control of an experiment or an instrument falls into two basic categories: open loop and closed loop. An open-loop control computer is simply telling the instrument or process to perform specific operations at specific times. We need not even use a computer for this function. The simplest case of an open-loop control system might be the automatic soft drink machine. You put your coin in the slot, push buttons and take out a drink; that is if the machine is functioning properly. How many times have you put money into a soft drink machine only to have the paper cup jam and to watch your soft drink squirt down the drain? This is, of course, the basic disadvantage of any open-loop control system. You can sequence the machine; that is, you can let it know what you want it to do, but you give the machine no alternative in case something goes wrong.

Closed loop control, on the other hand, allows the computer to make judgements. First, the machine is sequenced and then the computer examines the state of the machine to determine if it is performing properly. Based on the state of the instrument or process, the computer can make a decision on what the instrument or process should do next. A very interesting example of this is found in an astronomical observatory at a midwestern university. Here some scientists have computerized the star tracking system. Not only does the computer have the ability to start the system automatically once darkness has fallen, and to seek out stars, but it also has the decision making ability of not operating during cloudy or inclement weather. It also has the judgemental capability of waiting to see whether the weather will clear up before shutting itself down for good. To show how flexible this system is, when the automatic telescope was first tried out there was a malfunction in one of the drive motors of the tracking system. After attempting to initiate several starts the computer was able to realize the malfunction of the system and work around it so that the observations made that night were the ones desired. Here was the case where decision making by a machine allowed data to be recorded that otherwise would have been lost. Another example of closed loop control is the X-ray diffractometer. Here the computer drives the stepping motors of the goniometer and, based on measurements made at certain angles, the computer calculates other angles at which measurements should be made. Thus, the computer can reduce, from several days to several hours, the time needed to take measurements.

In addition to control, a computer is obviously capable of acquiring data. In the aforementioned star tracking and X-ray diffractometer systems the computer collects data which is analyzed off line; by an astronomer in one case, and by a much larger computer in the other case.

Dedicated computers are capable of much more than control and data acquisition however. It is quite routine nowadays to have the computer performing the data reduction and analysis for the experiment. An example of this would be an NMR spectrometer where the computer is time averaging a noisy signal, a high-resolution mass spectrometer where the computer is calculating the most probable elemental composition and mass mapping, and a gas chromatograph where the computer is calculating component concentrations.

## 5. ADVANTAGES

This would be a good time to examine the relative advantages of dedicated small computers and time sharing systems.

Perhaps the greatest advantage of the small dedicated computer is its low initial price. Typical systems will run in the $12–45,000 range. Because the computer is an integral part of the instrument, it is also very easy to use. Special languages have been developed so that, for example, a gas chromatographer may set up the parameters of his experiment through a question and answer conversation with a computer. He can use the dedicated computer as an analytical aid without ever having the necessity of becoming a programmer.

The dedicated small computer also means that it is often possible to have much more computer capability applied to a particular problem than when you are trying to interface to a time-sharing system which has other demands being made on it.

The dedicated computer also provides redundancy. For example, a laboratory having several chromatographs, an NMR, and a high-resolution mass spectrometer, with a computer dedicated to each one of these analytical instruments, will continue to operate in the event that one of the computers go down. It is important to note here that the cost of multiple dedicated computers is often much less than the cost of a time sharing system of equivalent capability.

With a dedicated computer users may also write and debug systems programs without interfering with other users. In medium-sized time-sharing systems, the computer is often so completely occupied that it is very difficult or impossible to modify and debug analysis routines while the computer is performing its time sharing function.

In addition, a dedicated computer system does not need large pro-graming staff nor does it need elaborate physical facilities complete with air conditioning. A PDP-8/1, for example, will run from 32 to 130°F and from 5 to 95% humidity. This means that the computer may be located right at the instrument, and not in a central compcenter.

Two of the greatest advantages, however, of the dedicated computers are the many turnkey systems which are available, and the ease with which these systems are expanded. It is now possible to buy dedicated computer systems for gas chromatography, NMR, high-resolution mass spectrometry, medium resolution mass spectrometry, X-ray diffractometry, X-ray spectro-metry, exhaust emission analysis, and emission spectroscopy. In many cases, these systems are sold by the manufacturers of the instruments themselves. They take the responsibility for making the physical connection between the computer and the instrument, and the responsibility for making the computer/ instrument combination operate. An additional advantage here is that the manufacturer also accepts responsibility for the maintenance and updating of the software. The availability of turnkey system means that there is no longer a long development between the time when the computer arrives on sight and the time at which it begins to satisfactorily perform its designed function.

Perhaps you are worrying about being closed in when purchasing a dedicated computer. As has been demonstrated time and time again this is very definitely not the case. By themselves, small computers are amazingly powerful analytical tools. Small dedicated computers can be gracefully expanded in one of two ways: by adding additional peripherals to the computer or by interfacing the dedicated computer to a larger computer or time sharing system.

Using the latter approach, a powerful central computer can service many dedicated computers simultaneously in a time sharing mode, greatly enhancing the capabilities of all of the systems. This is, in fact, one way in which a dedicated computer system may be expanded to a full-blown time-sharing system. It has the advantage of enabling the experimenter to start out slowly and build up the capability of his system without having to jump in with both feet at day one. The total system cost is comparable to that of most of the time sharing systems commercially available today.

Dedicated computers have gained very wide acceptance in the last five years. There are approximately 3500 small computers in the field today; many of which are being used as dedicated systems in conjunction with analytical instruments. Many prepackaged computer/software systems are available as "black box" devices for scientists who do not want to design their own systems. For the researcher who wishes to use the computer for other application, the dedicated computer offers ease of expandability and modification not often found in other types of computer systems.

# III. REAL-TIME HIGH-RESOLUTION MASS SPECTROMETRY

The Measurement of Accurate Molecular and Fragment Mass, and Relative Ionic Abundance: The Detection and Identification of Unresolved Isobaric Species.*

A. L. Burlingame, D. H. Smith, T. O. Merren,† and R. W. Olsen

*Space Sciences Laboratory, University of California*
*Berkeley, California*

A detailed description of the performance of a computer-coupled, high-resolution mass spectrometer system is presented. The results obtained from this system, which consist of accurate mass and intensity measurements for all ionic species present in a mass spectrum, are presented and discussed in the context of both system evaluation and application to organic analysis. The results indicate that mass measurement accuracies of 1 ppm or better are routinely obtainable through the use of the four-average technique, and that intensity measurement accuracy is limited only by ion statistics. Furthermore, system performance remains the same over a wide range of scan rates and mass spectrometer resolutions. Applications of such a system to organic analysis permits rapid acquisition and analysis of spectra and drastically reduces elemental composition ambiguities. A method for resolution of unresolved multiplets based on the mass measurement performance of the system is presented.

## 1. INTRODUCTION

The task of obtaining complete high-resolution mass spectra of complex organic molecules has involved use of computers for data processing for several years ([1]). The desirability of obtaining these spectra for structure elucidation or molecular fragmentation studies has been discussed ([2]). Our efforts have been directed toward developing more rapid and flexible

*Paper XXIII in the series High Resolution Mass Spectrometry in Molecular Structure Studies; for Part XXII, see A. L. Burlingame, D. H. Smith, and R. W. Olsen, *Anal. Chem.* (in preparation). Support of the National Aeronautics and Space Administration, Grants NGR 05-003-134, NsG 243, Suppl. 5, and NAS 9-7889, is gratefully acknowledged.

†Permanent address: G.E.C.–A.E.I., Ltd., Scientific Apparatus Division, Barton Dock Road, Urmston, Manchester, England.

means for acquiring such complete high-resolution mass spectra. This would be particularly suited for direct sample introduction employing capillary gas chromatographic techniques.

With the availability of improved performance, high-resolution mass spectrometers and high-speed digital computers, considerable interest has been expressed recently in linking these two devices in order to facilitate the task of data acquisition. These data are gathered while the experiment is in progress; thus the term "real-time." Several reports [2] have been presented on this subject, and a description of the performance of a proto-type system for collecting complete high-resolution mass spectral data employing a digital computer in real-time has appeared from this labora-tory [3]. This system produced mass measurements accurate to about 5 ppm and adequate abundance measurements. Subsequent improvements in the system involving use of a different mass spectrometer-computer system and updating of interfacing have been discussed [4]. These systems have proven to be quite flexible in terms of instrument operation, computer programing, and data display features, in addition to providing a rapid method for acquiring the vast amounts of data present in a high-resolution mass spectrum.

In this paper, we wish to discuss in some detail the quality of mass and intensity measurement data routinely obtainable from such a real time high resolution mass spectrometer. It was of considerable interest to perform a detailed evaluation of results obtained by this system. Such an evaluation provides a frame of reference in which system performance and limitations are well understood. Applications of such a computer-coupled high-resolution mass spectrometer system to routine problems in organic mass spectrometry can then proceed on a very firm footing.

These applications have led to development of a technique, based on mass measurement accuracy, for detection and identification of unresolved multiplets. Previous efforts directed toward the detection and identification of unresolved multiplets encountered in high-resolution mass spectra have concerned the mathematical deconvolution of the observed peak density profiles recorded on photographic plates [5,6]. Difficulties have arisen in attempting such treatment from many factors associated with the repro-ducibility of line images on the photographic plate, e.g., nonlinearity of emulsion blackening characteristics over a wide dynamic range in intensity, "noisy" peaks, and so forth. The technique discussed in this paper is quite insensitive to instrument parameters such as resolution and peak shape.

## 2. EXPERIMENTAL

A detailed description of the mass spectrometer computer system will appear elsewhere [7]. Briefly, the system includes an Associated Electrical

Industries, Ltd. modified MS-902 high-resolution mass spectrometer, capable of providing a resolution of 1 part in 50,000. Spectra are scanned by sweeping the magnetic field to cover a selected high-mass to low-mass range. The voltage output of an electron multiplier–amplifier system is digitized by an analog-to-digital (A/D) converter. The A/D output is linked to a Scientific Data Systems Sigma-7 computer. The output consists of peak voltage profiles recorded on magnetic tape and/or displayed on a cathode ray tube adjacent to a mass spectrometer. The tape is processed at a later time. Masses are calculated by relating the sample peak times to those of perfluorokerosene (PFK) through an exponential relationship of mass to time (Mass $\propto e^t$).

## 3. SYSTEM EVALUATION

The evaluation of data obtained by direct digitization on–line to the Sigma-7 will be concerned first with mass measurement accuracy and will delineate the performance which can be expected of the system described. Subsequently, relative abundance measurements will be evaluated and discussed. Before presenting the results, however, the method of calculation of the number of ions, N, in a peak will be discussed. The quantity N is used in evaluation of relative abundance measurements.

### 3.1. Calculation of *N*

The total number of ions may be estimated for each peak (singlet or multiplet) from the raw digital data. In practice, $N$ is calculated and output during data reduction on the Sigma-7. $N$ is calculated using the relationship derived below.

With reference to the peak depicted in Fig. 1, consider an element, of duration $t$ seconds, of a peak, the element containing $n$ ions.

The rate of arrival of $n$ ions is therefore $n/t$ per sec. If $e$ is the charge of an electron, then the current $i$ is given (in amperes) by

$$i = \frac{en}{t}$$

This ion current strikes the first dynode of an electron multiplier and is amplified by a factor $G$, the gain of the multiplier. On the last dynode, therefore, the current is

$$i = \frac{Gen}{t}$$

Attached to the last dynode is an amplifier, which has an input resistance $R$. The voltage $E$ generated for this increment is then given by

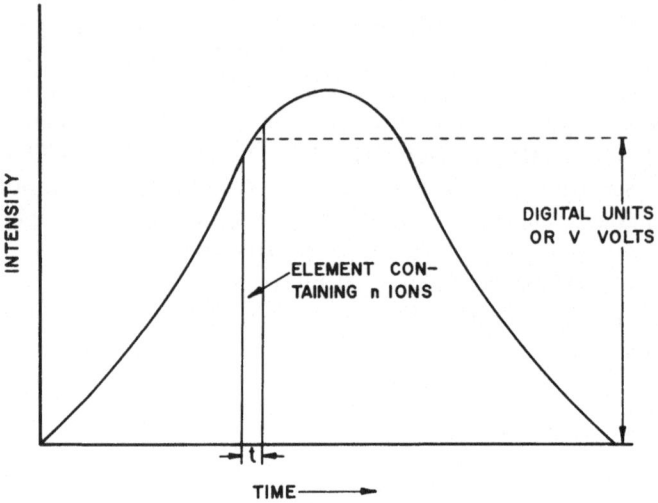

Figure 1. Illustration of parameters used in derivation of equation 1.

$$E = iR = \frac{RGen}{t}$$

The element of time considered $t$ is the time between successive digitizations by the A/D, so that

$$t = \frac{1}{f}$$

where $f$ is the digitization rate.

$$E = GRenf$$

Now, let the voltage level represented by one binary "bit" on the A/D converter be $v$ volts.* If one observes $c$ bits, then

$$E = cv$$
$$\therefore cv = GRenf$$

and

$$n = \frac{cv}{GRef}$$

The total number of ions is then given by a summation over all time elements per peak:

$$\Sigma n = N = \Sigma c \times \frac{v}{GRef} \tag{1}$$

*The output of the A/D is a binary number, from 0 to $2^{14} - 1$ since there are 14 bits plus sign for this particular A/D. Therefore, 0–10 V represents 0 bits to 16,383 bits.

The summation over $c$ is simply the peak area obtained by summation of the data points comprising a peak profile. The other parameters are noted before data are taken.

There is an additional consideration of importance, and that is the fact that the measured multiplier gain depends on the elemental composition of the particular ionic species impinging upon the electron multiplier ([8]). The gain is generally measured using a peak of PFK. It has been determined that the gain is 10–15% higher for the hydrocarbon or oxygen containing ions, depending on the particular elemental composition, than for the peaks of perfluorokerosene. It should be kept in mind, therefore, that the calculated number of ions represents an *upper* limit on $N$.

## 3.2. Mass Measurement Accuracy

In order to assess the accuracy of mass measurement attainable, nine successive scans were made at 25,000 resolution, with a scan rate of 35 sec/ decade in mass. The data were digitized at 24 kHz. Perchlorobutadiene was chosen for the scans as a compound in which the masses of the fragment ions are known unambiguously. The sample flow rate into the ion source from a heated glass inlet system was less than 10 ng/sec.

For each of the runs, the differences between the observed and true masses were calculated for all peaks. Figure 2a shows the distribution of errors in the resulting 266 accurate mass values covering the mass range 100–266 with intensities greater than 2% of the base peak. Histogram A in Fig. 2a illustrates the percentage of errors (absolute values are plotted) falling within the ranges 0–2, 2–4 ppm and so forth. It can be seen that in 70% of the measurements, the observed mass differs from the true mass by less than 2 ppm.

An improvement in accuracy has been obtained by calculating the mean mass for several scans. If the errors in individual observed masses are random, then both precision and mass accuracy should be improved by a factor of two, taking the mean mass in four scans, and a factor of three taking the mean mass for nine scans. If, however, the observed deviations are due to systematic errors in the data acquisition and/or data reduction systems, mass measurement accuracy taking the mean of several scans should not be significantly better than for single scans.

To evaluate the data, mean masses were calculated for scans 1–4 and 5–8 of the nine scans. The 58 differences obtained are plotted in Fig. 2a as histogram B. In this case, the ranges are 0–1, 1–2 ppm and so forth. Comparison with histogram A shows that the errors are reduced by about a factor of two, with 77% of the errors now less than 1 ppm.

The mean differences for all nine scans are shown in Fig. 2a as histogram C, plotted over the ranges 0.0–0.5, 0.5–1.0 ppm, and so forth. Again the

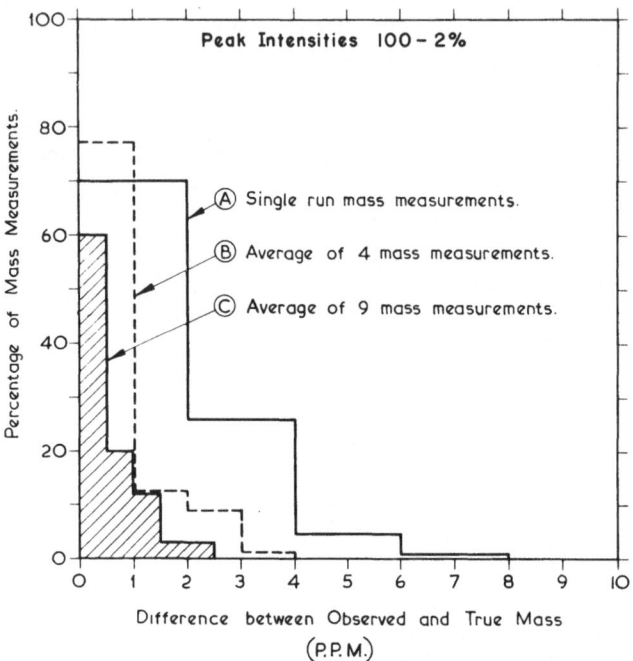

Figure 2a. Distribution of mass measurement errors at resolution of 25,000.

accuracy has been improved. In this case, 60% of the values are less than 0.5 ppm. These improvements are all consistent with observed mass measurement accuracy being a result of random rather than systematic errors.

Figure 2b shows that a further improvement in accuracy is obtained if only those peaks greater than 10% of the base peak are considered. The values plotted in the histograms were obtained exactly as for Fig. 2a. Seventy-eight percent of the differences are less than 2 ppm for single measurements, 88% are less than 1 ppm for the mean of four scans, and 71% are less than 0.5 ppm for the mean of nine scans. Furthermore, comparison of the histograms in Figs. 2ab shows that the larger errors are eliminated considering the smaller dynamic range.

The mass measurement accuracy described above is obtained for a total consumption of perchlorobutadiene of less than 0.5 $\mu$g in an individual scan and less than 2 $\mu$g for a group of four scans. Independent experiments on other organic compounds indicate that similar performance is obtained for similar sample quantities. The improved accuracy observed for peaks greater than 10% of the base peak in the above results would be expected to apply to all peaks down to 2% of the base peak if five times as much sample were used.

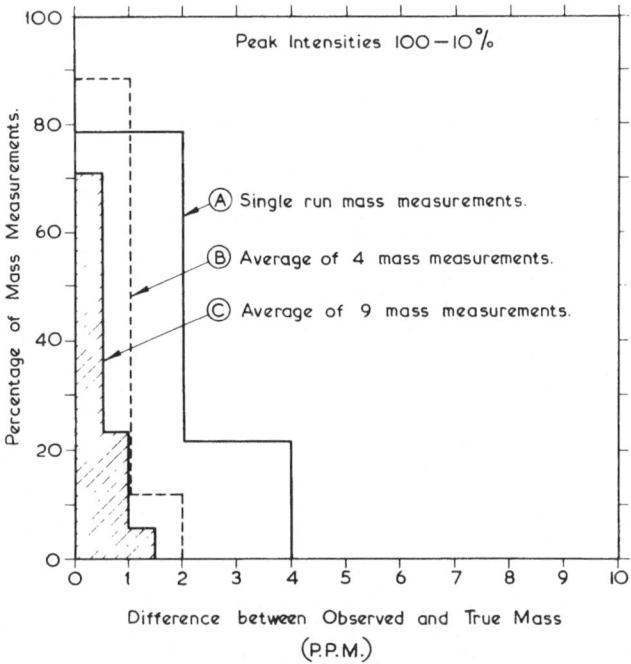

Figure 2b.  Distribution of mass measurement errors at resolution of 25,000.

Some representative observed errors are tabulated in Table I. Errors for a representative single scan, averages for scans 1–4 and averages for scans 1–9 are expressed in ppm and in millimass units (mmu). The results of scans 1–9 are those used for histogram C in Fig. 2b. The only value larger than 1 ppm in the 9 average column is that at $m/e$ 260, 1.15 ppm. The root-mean-square value is calculated for each column, since this value can be taken as a characteristic measure of performance for the particular number of scans considered.

The results indicate that a significant improvement in accuracy is achieved by taking four scans of a spectrum, and from both the sample and data handling points of view, the four-average method yields a practical improvement. Furthermore, the lack of systematic errors does enable mass measurement accuracy to be improved by multiple scans to about 0.5 ppm for peaks above 2% of the base peak.

The magnitudes of errors in millimass units for the above data provide an indication of the tolerances that must be considered for the acceptance or rejection of possible elemental compositions based on the observed mass. In Table I, for example, the R.M.S. value of 0.30 mmu with the largest observed error 0.91 mmu indicates the small tolerances that need be considered

**TABLE I.**

| Species | Nominal mass | Error, ppm | | | Error, mmu | | | Relative intensity |
|---------|--------------|------|-------|-------|------|-------|-------|-----------|
|         |              | 1    | (1–4) | (1–9) | 1    | (1–4) | (1–9) |           |
| $C_4Cl_2$ | 118 | −0.83 | −0.87 | −0.29 | −0.098 | −0.103 | −0.034 | 24.6 |
|           | 120 | −0.41 | 0.12  | 0.03  | −0.049 | 0.014  | 0.004  | 16.8 |
| $C_3Cl_3$ | 141 | −0.64 | −0.42 | −0.31 | −0.090 | −0.059 | −0.044 | 21.6 |
|           | 143 | −1.26 | 0.09  | 0.17  | −0.180 | 0.013  | 0.024  | 21.6 |
| $C_4Cl_3$ | 153 | −0.83 | 0.46  | 0.12  | −0.127 | 0.070  | 0.018  | 14.6 |
|           | 155 | 0.83  | 0.71  | 0.63  | 0.129  | 0.110  | 0.098  | 14.5 |
| $C_4Cl_4$ | 188 | −0.82 | −0.49 | −0.41 | −0.154 | −0.092 | −0.077 | 28.4 |
|           | 190 | −1.66 | −0.36 | −0.43 | −0.315 | −0.068 | −0.082 | 36.6 |
|           | 192 | 0.33  | 0.10  | −0.44 | 0.063  | 0.019  | −0.084 | 17.5 |
| $C_4Cl_5$ | 223 | −0.91 | 0.54  | 0.37  | −0.203 | 0.121  | 0.083  | 62.0 |
|           | 225 | −0.61 | −0.51 | −0.26 | −0.137 | −0.115 | −0.058 | 100.0 |
|           | 227 | 0.91  | −0.09 | 0.12  | 0.206  | −0.020 | 0.027  | 62.1 |
|           | 229 | 0.92  | 0.23  | 0.27  | 0.211  | 0.051  | 0.062  | 20.3 |
| $C_4Cl_6$ | 258 | 2.34  | 1.77  | 0.96  | 0.605  | 0.457  | 0.248  | 21.3 |
|           | 260 | 2.54  | 1.61  | 1.15  | 0.660  | 0.418  | 0.299  | 40.9 |
|           | 262 | 2.88  | 1.70  | 0.91  | 0.705  | 0.445  | 0.238  | 31.2 |
|           | 264 | 3.45  | 1.12  | 0.75  | 0.910  | 0.296  | 0.198  | 13.8 |
| | RMS VALUE | 1.59 | 0.859 ppm | 0.547 | 0.382 | 0.209 mmu | 0.132 | |

for only a single scan of a spectrum. Obtaining multiple scans yields the tabulated improvements in accuracy, and the resulting reduction of composition ambiguities possible. The R.M.S. value for nine scans is 0.13 mmu, with a largest error of 0.3 mmu.

Similar experiments were carried out at a resolution of 11,300 with a reduced sample flow rate (2.5 ng/sec). These results are presented as histograms in Figs. 3a and 3b as for the results at 25,000 resolution. The histograms show that the accuracy of mass measurement achieved at a resolution of 11,300 is very similar to that at a resolution of 25,000. For example, for the intensity range 100–2 %, Fig. 31, 73 % of the four-average mass measurements fall less than 1 ppm from the true value, and 63 % of the ten average measurements fall less than 0.5 ppm. Consideration of the intensity range 100–10 %, Fig. 3b, indicates results comparable to those shown in Fig. 2b.

At lower resolution, therefore, mass measurement accuracy is main-

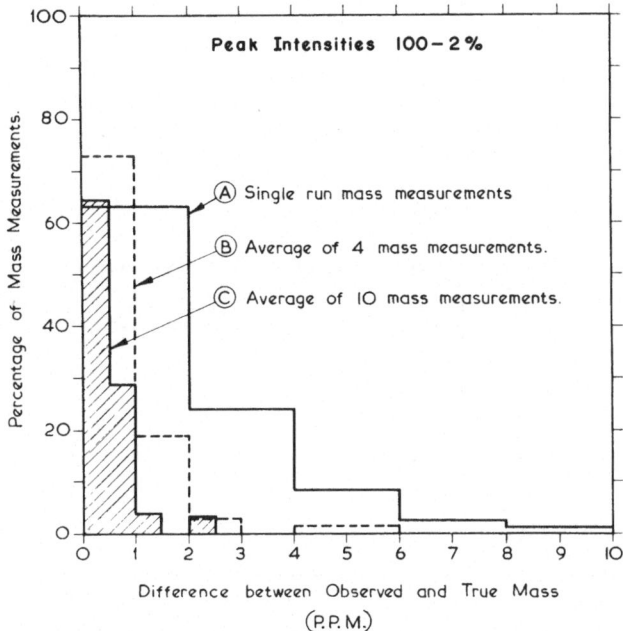

Figure 3a. Distribution of mass measurement errors at resolution of 11,000.

tained using smaller sample quantities. For this experiment, less than 0.5 $\mu$g flowed into the ion source for the period during which four scans were taken. Since Figs. 2a, 2b, 3a, and 3b demonstrate that improved accuracy is obtained on more intense peaks, a corresponding improvement would be obtained for peaks above 2% of the base peak at a higher sample flow rate.

Results of experiments on other organic samples indicate the results obtained over a dynamic range of 500–1. A representative example is a series of eight scans of the spectrum of methyl arachidate (discussed in a subsequent section), the methyl ester of $n$-$C_{20}$ acid, at a resolution of 25,000. For these data, 65% of mass measurements (averaged over eight scans) fall within 1 ppm of the true mass, considering peaks above 0.2% of the base peak, and masses above $m/e$ 100. Considering all masses above $m/e$ 40 over this dynamic range, 61% of mass measurements are within 1 ppm of the true mass. The largest error noted is that of $m/e$ 44 ($C_2$ $^{13}C_1H_7$), 0.22 mmu, or 5.0 ppm. The slight decrease in mass accuracy below $m/e$ 100 is relatively unimportant, because there are many fewer possible ambiguities in elemental composition.

### 3.3. Intensity Measurement Precision and Accuracy

Use of a chlorinated compound permits a detailed study of intensity measurement accuracy. The intensity ratios of peaks containing the same

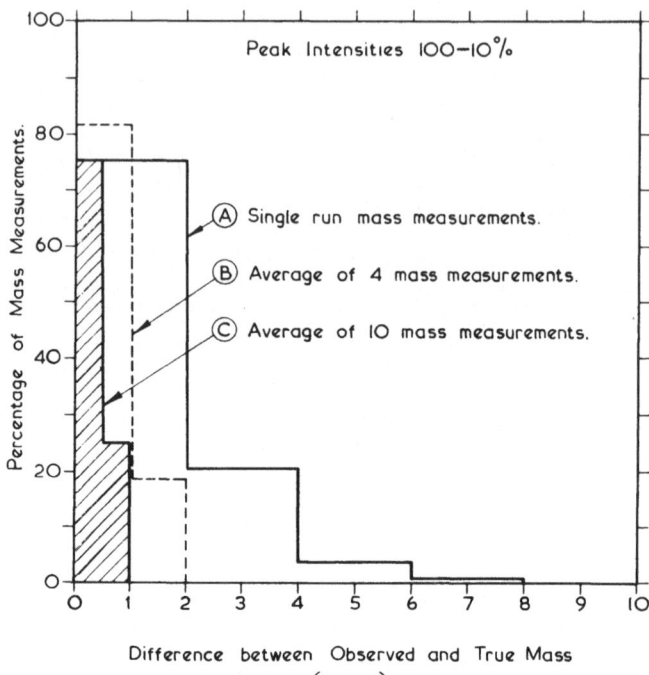

Figure 3b. Distribution of mass measurement errors at resolution of 11,000.

number of carbon and chlorine atoms can be calculated from the formula $(a + b)^m$ where $m$ is the number of chlorine atoms, and $a = 0.758$ and $b = 0.242$ are the natural abundances of $^{35}Cl$ and $^{37}Cl$. Intensity measurement precision may be evaluated by comparing the observed deviations in intensity with theoretical deviations based on ion statistics.

The evaluation of precision and accuracy can be accomplished as follows. Peak intensity measurements may be conveniently expressed as percentages of the total of isotopic peaks in a group, that is,

$$\frac{N_i}{\sum_i N_i} \times 100$$

where $N_i$ is the number of ions in peak $i$ (obtained as discussed in Section 1 above) and $\sum_i N_i$ is the total number of ions in the group.

The percentage abundances can be assigned theoretical standard deviations according to the equation

$$\sigma_{\text{theor}} = \frac{\sqrt{N_i}}{\sum_i N_i} \times 100$$

This form of the equation for $\sigma_{theor}$ results in the standard deviation expressed in the same units (percentage of total ions in the group) as the abundance value itself.

Nine successive scans of the spectrum of perchlorobutadiene at each of two resolutions, 11,300 and 30,000, were obtained, and these data used to evaluate intensity measurement. Representative data in Table II compare the mean observed abundances with the true abundances, and the observed standard deviations with the theoretical value.

Precision may be evaluated by comparing the observed and theoretical standard deviations. It can be seen that they agree very closely, indicating that variations in intensity measurements are fully accounted for by ion statistical considerations. This confirms the conclusions reached by McMurray et al. ([2]) on a limited number of peaks at 10,000 resolution, and extends them to a larger number of peaks, a wider intensity range and resolutions as high as 30,000.

Intensity measurement accuracy may be evaluated by comparing the observed and true abundances. A qualitative comparison indicates excellent agreement, judged in terms of the accuracy of intensity measurement desired in organic mass spectrometry. A quantitative assessment involves comparison with the expected standard deviation of the mean, that is, the theoretical standard deviation divided by $\sqrt{n}$ where $n$ is the number of scans from which the mean is calculated. Since $n$ is 9 for both sets of scans, the errors in the mean abundances should be compared with the theoretical standard deviations divided by 3 $(\sigma_{theor}/\sqrt{9})$.

It may be seen from the table that 76% of the errors are less than one-third of a theoretical standard deviation. One hundred percent of the errors are less than 2 $\sigma_{theor}/\sqrt{9}$. The expected percentages of the errors, due entirely to statistical factors, are 68% and 96%, respectively. This close agreement indicates that no significant systematic errors are present in the mass spectrometer or data acquisition and processing systems. As with the precision, therefore, accuracy of intensity measurement is shown to be limited by ion statistical considerations.

## 4. APPLICATIONS TO ORGANIC ANALYSIS

The results presented above were intended to provide a detailed evaluation of the mass spectrometer–computer system. With this evaluation in hand, it is now possible to examine results obtained on organic samples of interest for a mass spectroscopist. Specifically, it is of interest to examine how the conclusions arrived at in the preceding sections can be used for analysis of typical, everyday spectra.

One of the more important results observed above is that four repeated

## TABLE II.

| Species | m/e | True abundance | Resolution 11,300 | | | | Resolution 30,000 | | | |
|---|---|---|---|---|---|---|---|---|---|---|
| | | | Mean observed abundance | Standard deviation | Theoretical standard deviation | Error in mean abundance | Mean observed abundance | Standard deviation | Theoretical standard deviation | Error in mean abundance |
| $C_4Cl_6$ | 258 | 19.0 | 18.8 | ±0.5 | 0.5 | -0.2 | 18.8 | ±1.3 | 1.5 | -0.2 |
| | 260 | 36.3 | 36.5 | ±0.4 | 0.7 | +0.2 | 36.6 | ±2.2 | 2.1 | +0.3 |
| | 262 | 29.0 | 28.9 | ±0.8 | 0.7 | -0.1 | 28.7 | ±1.6 | 1.9 | -0.3 |
| | 264 | 12.4 | 12.4 | ±0.3 | 0.4 | 0.0 | 12.6 | ±0.9 | 1.2 | +0.2 |
| | 266 | 2.96 | 3.00 | ±0.17 | 0.22 | +0.04 | 3.12 | ±0.41 | 0.61 | +0.16 |
| | 268 | 0.38 | .35 | ±0.12 | 0.07 | -0.03 | 0.38 | ±0.20 | 0.22 | 0.00 |
| | 270 | .02 | — | — | — | — | — | — | — | — |
| $C_4Cl_5$ | 223 | 25.0 | 25.0 | ±0.5 | 0.4 | 0.0 | 24.5 | ±1.5 | 1.2 | -0.5 |
| | 225 | 39.9 | 39.6 | ±0.5 | 0.5 | -0.3 | 40.2 | ±1.6 | 1.5 | +0.3 |
| | 227 | 25.6 | 25.7 | ±0.4 | 0.4 | +0.1 | 25.8 | ±0.9 | 1.2 | +0.2 |
| | 229 | 8.16 | 8.31 | ±0.20 | 0.24 | +0.15 | 8.24 | ±0.54 | 0.68 | +0.08 |
| | 231 | 1.32 | 1.34 | ±0.14 | 0.09 | +0.02 | 1.25 | ±0.15 | 0.26 | -0.07 |
| | 233 | .08 | — | — | — | — | — | — | — | — |
| $C_4Cl_4$ | 188 | 33.0 | 33.1 | ±0.7 | 0.8 | +0.1 | 32.7 | ±1.6 | 2.3 | -0.3 |
| | 190 | 42.2 | 42.3 | ±0.5 | 0.8 | +0.1 | 42.2 | ±2.5 | 2.6 | 0.0 |
| | 192 | 20.2 | 20.3 | ±0.8 | 0.6 | +0.1 | 20.5 | ±2.3 | 1.8 | +0.3 |
| | 194 | 4.31 | 4.35 | ±0.27 | 0.28 | +0.04 | 4.56 | ±0.72 | 0.85 | +0.25 |
| | 196 | 0.30 | — | — | — | — | — | — | — | — |
| $C_4Cl_3$ | 141 | 43.6 | 43.7 | ±1.1 | 1.1 | +0.1 | 43.8 | ±5.1 | 3.3 | +0.2 |
| | 143 | 41.7 | 41.6 | ±1.1 | 1.1 | -0.1 | 42.3 | ±2.3 | 3.3 | +0.6 |
| | 145 | 13.3 | 13.3 | ±0.5 | 0.6 | 0.0 | 12.8 | ±1.9 | 1.8 | -0.5 |
| | 147 | 1.42 | 1.38 | ±0.28 | 0.20 | -0.04 | 1.10 | ±0.53 | 0.57 | -0.32 |
| $C_4Cl_2$ | 118 | 57.5 | 56.8 | ±1.3 | 1.3 | -0.7 | 57.7 | ±3.6 | 4.3 | +0.2 |
| | 120 | 36.7 | 37.4 | ±1.1 | 1.1 | +0.7 | 36.0 | ±3.9 | 3.4 | -0.7 |
| | 122 | 5.80 | 5.81 | ±0.48 | 0.42 | +0.01 | 6.30 | ±1.37 | 1.41 | +0.50 |

scans of a spectrum will yield mass measurement accuracies of between one and two ppm. It may be expected that in most situations it is possible to obtain four spectra of a compound. To illustrate the information that may be obtained by knowing the mass measurement errors expected from the system, eight successive scans of the spectrum of methyl arachidate ($n$-$C_{19}H_{39}COOCH_3$) were recorded at 10,000 resolving power.* Mass measurement accuracy was evaluated by calculating average mass differences (in ppm), between assigned and observed mass, for spectra 1–4, 5–8, and 1–8. Some data representative of those observed for the complete spectrum are presented in Table III.

An examination of Table III reveals two sets of measurement accuracy data. One set, the majority of the table, includes those data that compare favorably with what is expected from the previous sections. The other set, comprising nominal masses 70, 88, 130, 186, and 214, include those data

**TABLE III. Mass Measurement Accuracy. Methyl Arachidate**

| Assigned composition | Assigned mass | Average difference, ppm | | | Average intensity |
|---|---|---|---|---|---|
| | | 1–4 | 5–8 | 1–8 | |
| $C_3H_5$ | 41.03912 | −0.51 | −1.62 | −1.06 | 7.81 |
| $C_2{}^{13}C_1H_4$ | 42.04247 | −0.48 | −1.97 | −1.32 | 0.26 |
| $C_3H_6$ | 42.04694 | −1.03 | 0.26 | −0.39 | 2.49 |
| $C_2H_3O_2$ | 59.01330 | 1.32 | 0.85 | 1.08 | 1.43 |
| $C_5H_9$ | 69.07042 | −0.53 | −0.37 | −0.45 | 12.45 |
| $C_5H_{10}$ | 70.07824 | −11.86 | −12.53 | −12.20 | 3.50 |
| $C_3H_6O_2$ | 74.03677 | 0.27 | −0.95 | −0.34 | 100.00 |
| $C_4H_7O_2$ | 87.04459 | 0.43 | 0.68 | 0.50 | 65.14 |
| $C_4H_8O_2$ | 88.05242 | −27.48 | −25.30 | −26.39 | 5.51 |
| $C_5H_9O_2$ | 101.06024 | −0.35 | 0.67 | 0.16 | 6.81 |
| $C_9H_{13}$ | 121.10172 | −0.76 | −1.55 | −1.15 | 0.98 |
| $C_7H_{13}O_2$ | 129.09155 | −1.52 | −0.38 | −0.95 | 7.19 |
| $C_7H_{14}O_2$ | 130.09936 | −9.03 | −7.87 | −8.45 | 2.30 |
| $C_8H_{15}O_2$ | 143.10719 | 0.25 | −0.37 | −0.06 | 19.84 |
| $C_{12}H_{19}$ | 163.14866 | −0.18 | 0.82 | 0.32 | 0.39 |
| $C_{11}H_{22}O_2$ | 185.15414 | 0.28 | −0.09 | 0.10 | 4.93 |
| $C_{11}H_{23}O_2$ | 186.16196 | −12.79 | −12.21 | −12.45 | 1.13 |
| $C_{13}H_{25}O_2$ | 213.18544 | 0.62 | −0.13 | 0.25 | 2.05 |
| $C_{13}H_{26}O_2$ | 214.19326 | −10.05 | −8.68 | −9.38 | 0.70 |
| $C_{18}H_{34}$ | 250.26603 | −0.48 | 1.17 | 0.25 | 0.18 |
| $C_{18}H_{35}O_2$ | 283.26368 | 0.25 | −0.46 | −0.10 | 8.82 |
| $C_{19}H_{37}O_2$ | 297.27933 | −0.51 | 0.47 | −0.02 | 1.82 |
| $C_{21}H_{42}O_2$ | 326.31846 | −0.28 | −0.63 | −0.45 | 15.34 |

*Spectra scanned at 35 sec/decade with a digitization rate of 24 kHz.

that exhibit very high differences. These entries may then represent unresolved doublets, with the contribution from the second peak shifting the peak position sufficiently to yield errors higher than expected.

To illustrate this situation in more detail, peak profiles from one scan of the resolved $^{13}$C *vs.* CH doublet at $m/e$ 42 ($\Delta M = 0.00446$ amu) and the suspected unresolved doublets at $m/e$ 70, 88, and 130 are presented in Figs. 4 and 5. Peak widths at 10,000 resolution, 35 sec/decade, should be $\sim 1500\,\mu$sec. At a digitization rate of 24 kHz, the peaks should be $\sim 36$ clock pulses wide at the 5% level. Calculated positions of the center of gravity (labeled C.G. in the figures) are indicated. It is observed that the profiles of $m/e$ 70 and 88 do indeed indicate unresolved doublets. These doublets, as will be described below, are due to $^{13}$C *vs.* CH. The presence of an unresolved $^{13}$C isotope peak under the peak profile should result in a shift of the center of gravity to lower mass, resulting in a negative difference when compared to the mass of an assigned composition including CH, rather than $^{13}$C. This is exactly what is observed in Table III (note the large negative differences of $m/e$ 70, 88, 120, 186, and 214). The profile of $m/e$ 130 (Fig. 5) which is only slightly suggestive of an unresolved doublet, is a particularly striking example of how little the center of gravity need be shifted to yield a large mass measurement error.

Further investigations were carried out by recording repetitive scans

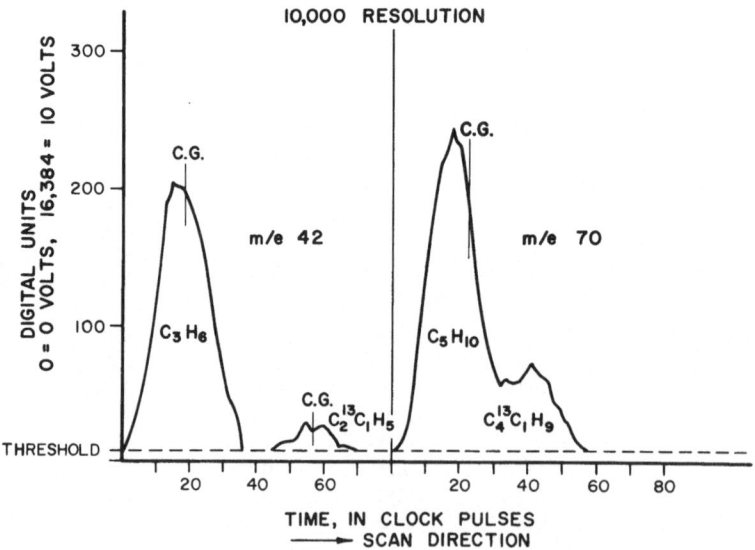

Figure 4. Peak profiles of $m/e$ 42 and $m/e$ 70 at a resolution of 10,000. Scan rate 35 sec/decade in mass. Clock rate 24 kHz. Mass spectrum of methyl arachidate.

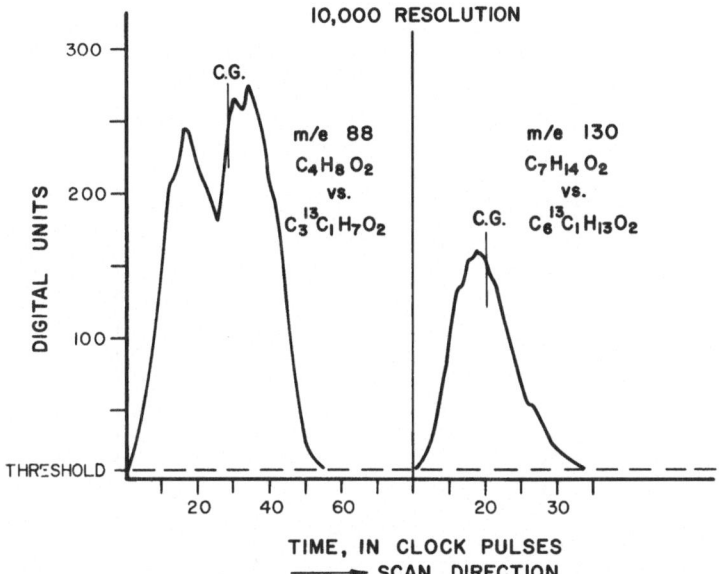

Figure 5. Peak profiles of $m/e$ 88 and $m/e$ 130 at a resolution of 10,000. Scan rate 35 sec/decade in mass. Clock rate 24 kHz. Mass spectrum of methyl arachidate.

at higher resolving powers.* These results are tabulated in Table IV for some of the suspected doublets noted at 10,000 resolving power. These data are representative of the fact that, for these spectra, masses having differences greater than expected on the basis of system evaluation are indeed unresolved doublets of the type CH $vs.$ $^{13}$C.

Mass measurement thus provides a criterion for the detection of unresolved doublets. It is important to point out, however, that high errors can also result from not considering, in determination of elemental composition and thus assigned mass, sufficient numbers and/or kinds of heteroatoms that may be present in the sample. With the condition that the numbers and kinds of heteroatoms present are known, it is possible to detect unresolved doublets in a spectrum. Another consideration of importance is an estimation of how large mass measurement errors must be to detect these doublets. This involves not only the *mass* separation of the peaks, but also the ratio of their intensities. It is possible to derive a relation between these variables to determine the range of intensity ratio and mass separation to yield a given mass measurement error.

With reference to Fig. 6, consider a doublet comprised of peaks of intensities (areas) $I_1$ and $I_2$, with separation $S$. Let the center of mass of

*25,000 resolution scans taken at 35 sec/decade; digitization rate of 24 kHz.
 30,000 resolution scans taken at 64 sec/decade; digitization rate of 24 kHz.

**TABLE IV. Doublet Recognition**

| Nominal mass | Assigned composition | Resolution | | | | | |
|---|---|---|---|---|---|---|---|
| | | 10,000 (8 scans) | | 25,000 (9 scans) | | 30,000 (6 scans) | |
| | | Difference* | Rel. int., % | Difference | Rel. int., % | Difference | Rel. int., % |
| 70 | $C_5H_{10}$ | −12.20 (U)† | 3.50 | 0.67 | 4.10 | 0.43 | 4.32 |
| | $C_4{}^{13}C_1H_9$ | | | 1.03 | 0.85 | −1.06 | 0.67 |
| 88 | $C_4H_8O_2$ | −26.39 (U) | 5.51 | −0.63 | 2.58 | 1.57 | 2.52 |
| | $C_3{}^{13}CH_7O_2$ | | | −0.25 | 3.17 | 0.69 | 2.61 |
| 130 | $C_7H_{14}O_2$ | −8.45 (U) | 2.30 | −0.67 | 2.12 | 0.63 | 2.10 |
| | $C_6{}^{13}C_1H_{13}O_2$ | | | 0.52 | 0.55 | 0.21 | 0.45 |
| 186 | $C_{11}H_{22}O_2$ | −12.45 (U) | 1.13 | −13.90 (U) | 1.08 | 0.09 | 0.60 |
| | $C_{10}{}^{13}C_1H_{21}O_2$ | | | | | −1.12 | 0.38 |
| 214 | $C_{13}H_{26}O_2$ | −9.38 (U) | 0.70 | −10.39 (U) | 0.47 | −0.47 | 0.20 |
| | $C_{12}{}^{13}C_1H_{25}O_2$ | | | | | 0.97 | 0.24 |

*Difference, in ppm, of average and assigned mass.
†'U' indicates unresolved doublet.

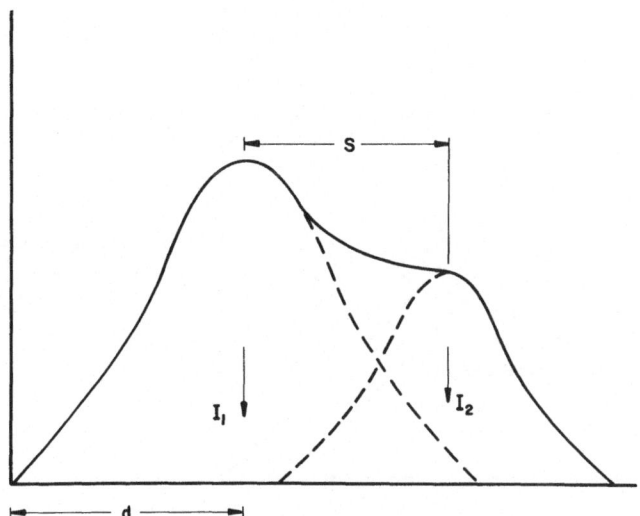

Figure 6. Illustration of parameters used in derivation of equation 2.

peak $I_1$ be $d$ from the origin. The calculated position of the doublet is given by:

$$\frac{I_1 + (d + S)I_2}{I_1 + I_2}$$

The shift $E$ of the larger peak is given by:

$$E = \frac{I_1 d + (d + S)I_2}{I_1 + I_2} - d$$

$$= \frac{I_2 S}{I_1 + I_2} \qquad (2)$$

$E$, which may be regarded as the error, may be expressed in ppm by expressing the mass separation $S$ as $\Delta M/M$, the difference in mass divided by the mass, time $10^6$. The value $S$ does *not* depend on the actual physical separation of the peaks. It is merely the separation of the centers of gravity of the two peaks comprising the doublet. Thus equation (2) is *independent* of mass spectrometer resolution.

The equation is plotted in Fig. 7 for the two error confidence limits of 1 and 2 ppm. Doublets, with appropriate separation $S$ and intensity ratios, falling in the hatched region should be detectable on the basis of mass measurement error at the indicated confidence level. For example, if one is confident that a series of mass measurements should yield an accuracy of better than 1 ppm for single, resolved peaks, then doublets with a separation $\Delta M/M = 4$ ppm are detectable if $I_2/I_1 \geq 35\%$, and doublets with $\Delta M/M = 20$ ppm are detectable if $I_2/I_1 \geq 5\%$.

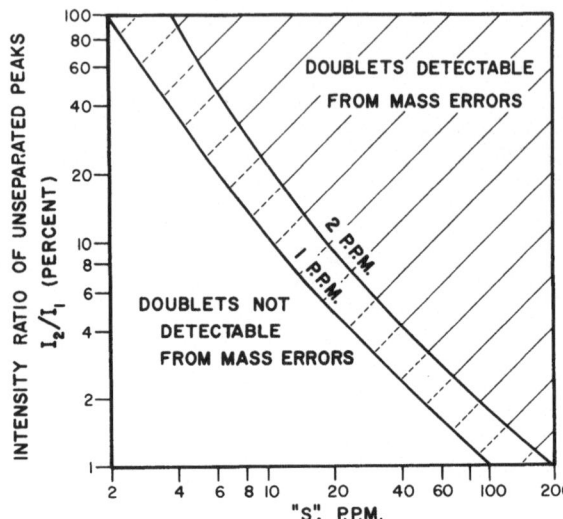

Figure 7. Intensity ratio of unseparated peaks *vs.* the separation $S = \Delta M/M$ for the error confidence limits of 1 and 2 ppm.

Equation (2) may be applied and tested for the CH *vs.* $^{13}C$ doublets noted above. Intensity $I_2$, the $^{13}C$ contribution, may be calculated on the basis of the corresponding peak one mass unit lower. $S$ and $(I_1 + I_2)$ are known quantities. Subtraction of the calculated error from the observed error yields a mass measurement error for the CH species. As may be seen from the results tabulated in Table V, this procedure works quite well.

**TABLE V.**

| Nominal mass | Observed error, ppm | Mass measurement error, CH species, ppm |
|:---:|:---:|:---:|
| 70 | −12.20 | 0.2 |
| 88 | −26.39 | −0.1 |
| 130 | −8.45 | −0.3 |
| 186 | −12.45 | −0.3 |
| 214 | −9.38 | −0.8 |

Thus, one obvious use of equation (2) is immediately suggested. This equation could readily be incorporated into a data reduction scheme to eliminate $^{13}C$ isotopic contributions to unresolved doublets, be they $^{13}C$ *vs.* CH or $^{13}C$ *vs.* another species, resulting in a new measured mass to match with a composition.

The above discussion indicates that a detailed knowledge of the mass measurement capabilities of the system provides one criterion, but not necessarily the only one, to screen the real time data for suspected unresolved multiplets. Judicious use of equation (2), based on knowledge of the elemental composition of the intact molecule and the calculated mass of the multiplet from the center of gravity, then allows calculation of the accurate mass and intensity of the peaks comprising the multiplet. Equation (2) may be applied in an iterative fashion for multiplets of higher order than doublets. Indeed, this method has been applied successfully to several triplets of the type $^{13}CH$ *vs.* $CH_2$ *vs.* CD ([9]). This simple mathematical procedure lends itself readily to automatic calculation by appropriate computer programing, and this method is currently being pursued in more detail in this laboratory.

This technique of multiplet resolution appears to offer distinct advantages over the technique of deconvolution in applications to electrically detected mass spectra. In addition to being mathematically much simpler to apply, a distinct advantage to those not having access to a large computer, multiplet resolution based on only two pieces of data, the accurate mass and intensity, eliminates many of the problems, such as peak profile smoothing and introduction of several empirical parameters, associated with application of deconvolution techniques ([5,6]).

For further applications to organic analysis, it is perhaps desirable to examine some of the results obtained on spectra of compounds with molecular weights considerably higher than those of perchlorobutadiene and methyl arachidate, discussed previously. These spectra indicate that mass measurement accuracy is maintained at the higher masses. Two examples will be discussed briefly.

The first example is concerned with spectra obtained on several steroid derivatives related to lanosterol. Mass measurements on one member of the series serve as an illustration. Four successive spectra were obtained on a sample of dihydrolanosterolenedione acetate (I) on the direct introduction probe.* Table VI presents some of the more prominent peaks in the spectrum above mass 350, with assigned mass and composition and average difference in ppm indicated.

(I)

---

*Spectra taken at 10,000 resolution, 24 kHz digitization rate.

**TABLE VI. Dihydrolanosterolenedione Acetate—High Mass**

| Assigned mass | Assigned composition | Average diff., ppm |
|---|---|---|
| 356.27151 | $C_{24}H_{36}O_2$ | −0.39 |
| 372.23004 | $C_{23}H_{32}O_4$ | 1.10 |
| 385.23787 | $C_{24}H_{33}O_4$ | −0.16 |
| 410.35485 | $C_{29}H_{46}O_1$ | 1.36 |
| 420.33920 | $C_{30}H_{44}O_1$ | 1.76 |
| 423.32629 | $C_{29}H_{43}O_2$ | 1.20 |
| 438.34976 | $C_{30}H_{46}O_2$ | 0.50 |
| 456.36032 | $C_{30}H_{48}O_3$ | 0.75 |
| 470.37597 | $C_{31}H_{50}O_3$ | 1.04 |
| 483.34741 | $C_{31}H_{47}O_4$ | 1.16 |
| 498.37089 | $C_{32}H_{50}O_4$ (M$^+$) | 2.18 |

The second example involves the spectrum of Rifamycin S (II), a member of a new series of antibiotics.*

(II)

Some representative data (single-spectrum mass measurements) for Rifamycin S are tabulated in Table VII.

**TABLE VII.**

| Species | Observed mass | Assigned mass | Composition | Difference, mmu | Relative intensity, % |
|---|---|---|---|---|---|
| M$^+$ | 695.29448 | 695.29415 | $C_{37}H_{45}NO_{12}$ | 0.33 | 1.53 |
| M—CH$_2$O | 665.28346 | 665.28359 | $C_{36}H_{43}NO_{11}$ | −0.13 | 1.46 |
| M—CH$_3$OH | 663.27001 | 663.26794 | $C_{36}H_{41}NO_{11}$ | 2.07 | 0.99 |

*We wish to thank Professor Sensi and Dr. Lancini for providing us with this sample.

The data in Tables VI and VII indicate that generalizations based on system evaluation at lower masses are extendable to higher masses. Mass measurement accuracy is maintained at these higher masses.

## CONCLUSIONS

A detailed examination of the technique of real-time high resolution mass spectrometry has been carried out. Evaluation of the system shows root mean square mass measurement accuracies of 0.5–3.0 ppm depending on the number of scans taken and the dynamic range covered by the data. The utility of the "four average" technique to provide accuracies sufficient to drastically limit elemental composition ambiguities has been described. Intensity measurement precision and accuracy has been shown to be limited only by ion statistical considerations.

This examination permits one to analyze data from organic samples of interest on the basis of known mass and intensity measurement performance of the system. Development of an equation to predict or analyze mass measurement errors greater than known limitations of the system has been described, and its application to doublets involving $^{13}C$ isotopic contributions has been illustrated. Data presented indicate retention of mass measurement accuracy to much higher masses than studied in the section on system evaluation.

It is felt that the data presented above provide an indication of the future potential of utilization of real-time computers in the field of high-resolution mass spectrometry.

## REFERENCES

1. P. Bommer, W. J. McMurray, and K. Biemann, 12th Annual Conference on Mass Spectrometry and Allied Topics, Montreal (June 7–12, 1964), p. 428.
   A. L. Burlingame, EUCHEM Conference on Mass Spectrometry, Sarlât, France, September 7–12, 1965.
2. A. L. Burlingame, *Advances in Mass Spectrometry, Vol. 4* (E. Kendrick, ed.), The Institute of Petroleum, London (1968), p. 15.
   W. J. McMurray, S. R. Lipsky, and B. N. Green, *ibid.*, p. 77.
   C. Merritt, Jr., P. Issenberg, and M. L. Bazinet, *ibid.*, p. 55.
   H. C. Bowen, E. Clayton, D. J. Shields, and H. M. Stanier, *ibid.*, p. 257.
   H. C. Bowen, T. Chenerix-Trench, S. D. Drackley, R. C. Faust, and R. H. Saunders, *J. Sci. Instrum.* **44**, 343(1967).
   W. J. McMurray, B. N. Green, and S. R. Lipsky, *Anal. Chem.* **38**, 1194(1966).
3. A. L. Burlingame, D. H. Smith, and R. W. Olsen, *Anal. Chem.* **40**, 13(1968).
4. A. L. Burlingame, D. H. Smith, R. W. Olsen, and T. O. Merren, 16th Annual Conference on Mass Spectrometry and Allied Topics, Pittsburgh (May 12–17, 1968).
   D. H. Smith, R. W. Olsen, and A. L. Burlingame, *ibid.*
5. D. D. Tunnicliff and P. A. Wadsworth, *private communication.*

6. R. Venkataraghavan, F. W. McLafferty, and J. W. Amy, *Anal. Chem.* **39**, 178(1967).
7. A. L. Burlingame, D. H. Smith, and R. W. Olsen, *Anal. Chem.* (*in preparation*).
8. C. La Lau, Mass discrimination caused by electron-multiplier detectors, *in: Topics in Organic Mass Spectrometry* (A. L. Burlingame, ed.), Wiley–Interscience, New York (in press).
9. A. L. Burlingame and D. H. Smith, unpublished results.

# IV. A COMPUTER CONTROLLED X-RAY SPECTROGRAPH

## Paul A. Weyler

*Research & Development—Extractive Metallurgy*
*Climax Molybdenum Company*
*Mines Park, Golden, Colorado*

One of the more recent developments in X-ray fluorescence spectrography is the use of a general-purpose digital computer to automatically operate the spectrometer and process the data for elemental analysis. This paper describes the various operating parameters controlled by the computer including such options as count-limit and repetition-limit and also the general mode of operation.

## 1. INTRODUCTION

The quest for automatic analysis of various materials by X-ray fluorescence spectrography has led to such developments as the "peg-board" type of programing and since then, during the past three years, to the use of a general-purpose digital computer. Both types set the operating parameters for each determination and provide a printed read-out. The extra knowledge gained by the use of a computer is that it will also process the data.

Since this paper is being presented at a session where there are mixed disciples of instrumental analytical chemists, I would like to review, very briefly, some of the features of X-ray fluorescence spectrography so that a better understanding of what has to be done "mechanically" and "theoretically" will develop.

As most probably many of you know, Bragg's equation is the basis of X-ray work:

$$n\lambda = 2d \sin \theta$$

where $n$ is the order of reflection; $\lambda$ is the wave length of the radiation; $d$ is the interplaner spacing of the crystal; and $\theta$ is the angle of reflection.

In diffraction work $d$ is the unknown, while in fluorescence $\lambda$ is unknown for qualitative work and the intensity of $\lambda$ is the variable for quantitative work.

Although Bragg's equation is quite simple, the equipment necessary to measure or satisfy the equation can be very sophisticated and, of course, expensive.

## 2. EQUIPMENT

The type of equipment that will be described in this paper (Fig. 1) is for a sequential operation. That is, one element (or one $2\theta$) is measured at a time. Simultaneous types would measure a prefixed number of elements at the same time.

The equipment consists of a Siemens Kristalloflex IV constant potential generator (60 kV–80 mA), Universal sequential spectrometer (SRS-1) with eight position sample changer, high-intensity tubes (Cr-2400W, Au-2800W), pulse spectroscope, and a circuit panel with two pulse height analyzer (PHA) window pots. The X-ray instruments are interfaced to a Digital Equipment Corporation Standard Programed Data Processor-8. The PDP-8 is designed for use as a small scale, general-purpose computer and is a one-address, parallel type with 12-bit, 4096 word core memory.

Figure 1. Equipment: From left to right—Kristolloflex IV generator, pulse spectroscope, SRS-1 with 8 sample changer, circuit panel, teletype typewriter, and PDP-8 computer.

The cycle time is 1.5 $\mu$sec and standard input–output facilities include teletype typewriter and perforated tape equipment.

## 3. OPERATING VARIABLES

A compilation of the operating variables or parameters of an X-ray unit is shown in Fig. 2. All the components could be controlled and operated automatically by the computer, but then each component would need a motor and computer device selector modules and cost and rationale must be maintained somehow. The asterisk denotes those variables controlled by this computer program.

### 3.1. X-ray Tube

1. Target: Usually two different tubes are kept on hand. Chromium target for the light elements and either tungsten, gold or platinum for the heavier elements. The choice would depend on the particular application.
2. Voltage and Amperage: Increasing voltage determines quality of the primary radiation and increasing amperage determines quantity of the primary radiation.

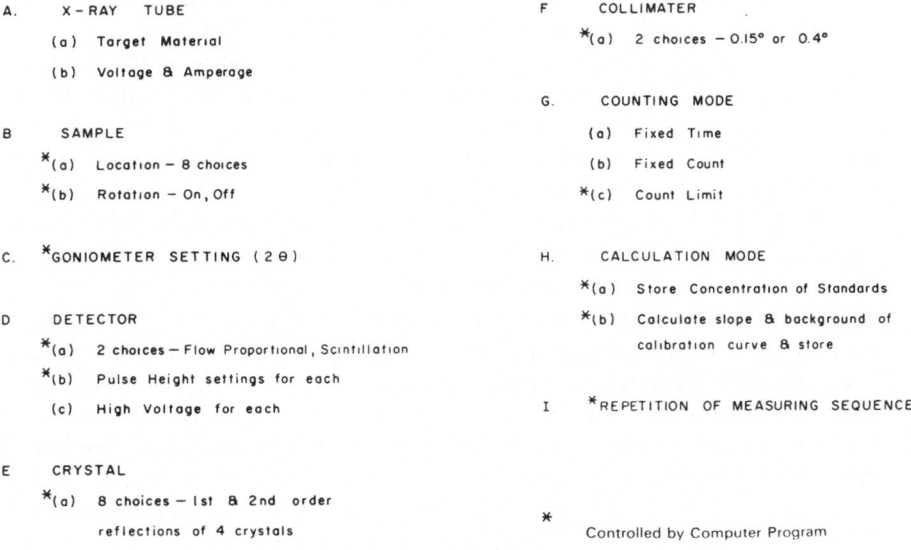

A.      X - RAY    TUBE

(a)   Target Material

(b)   Voltage & Amperage

B      SAMPLE

*(a)   Location – 8 choices

*(b)   Rotation – On, Off

C.   *GONIOMETER SETTING (2θ)

D      DETECTOR

*(a)   2 choices – Flow Proportional, Scintillation

*(b)   Pulse Height settings for each

(c)   High Voltage for each

E      CRYSTAL

*(a)   8 choices – 1st & 2nd order

reflections of 4 crystals

F      COLLIMATER

*(a)   2 choices – 0.15° or 0.4°

G.      COUNTING MODE

(a)   Fixed Time

(b)   Fixed Count

*(c)   Count Limit

H.      CALCULATION MODE

*(a)   Store Concentration of Standards

*(b)   Calculate slope & background of calibration curve & store

I      *REPETITION OF MEASURING SEQUENCE

*        Controlled by Computer Program

Figure 2. X-ray operating variables.

## 3.2. Sample

1. Location: Eight sample locations are provided for. The original SRS-1 is equipped for only two samples and our preference is to measure one element at a time so that it would be foolish in this case to have an automatic unit which required somebody to sit there and change samples between measurements. However, in the eight sample changer, locations $\theta$ and 1 must contain the standards for at least one set of measurements and any other samples must follow directly behind sequentially. There should be no skipping of samples in the holders.

2. Rotation: Some samples, regardless of refined sample preparation techniques, will show a preferred orientation which can only be compensated for by rotating the sample. One case is the $MoS_2$ concentrate at Climax where rotation is a must.

## 3.3. Goniometer Setting ($2\theta$)

Interface devices to control a function such as changing collimators, detectors, etc. (an either/or function) are quite readily adapted. However, to control and insure the goniometer setting is slightly more complicated.

The goniometer, at fast speed, is slewed at 300° $2\theta$/min and every revolution of the shaft is 1° $2\theta$. An incremental encoder is directly coupled to the goniometer shaft and gives 1000 pulses/rev of the shaft. Also, the encoder supplies positive, negative or zero pulses. Logically, the positive pulses are for increasing $2\theta$ and negative pulses for decreasing $2\theta$. In addition to this is a cam operated microswitch initially set to activate at some set angle, for example, 45°. The point at which the microswitch is activated and the goniometer stops is called the reference angle.

To describe this more fully, let us say that we want to start a measuring sequence. First, the $2\theta$ should be checked with the computer contents. A command can be given the computer and the $2\theta$ value in the computer will be typed out. If this varies from the $2\theta$ value on the goniometer, the goniometer should be changed manually to coincide with the computer value. After switching back to automatic again, an equal sign is typed. The goniometer will slow to 43° and start increasing $2\theta$ at a slow speed. As soon as the cam trips the microswitch, a zero pulse is sent to the computer, the goniometer is stopped and the $2\theta$ value is typed out. This value should be the same as the goniometer and is the starting point for the first sample change. After every eight measures, an automatic angle check will be made to insure the operator that the $2\theta$ is the proper one and that there is no mechanical error.

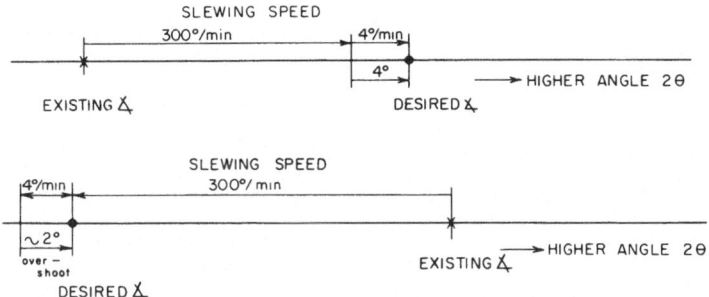

Figure 3. Goniometer control ($2\theta$ setting).

Figure 3 shows a change of $2\theta$ to a higher angle and to a lower angle. Note that the desired angle is always approached from the same side to prevent any backlash errors. The total number of degrees that the goniometer is operating at slow speed may be changed in the program.

## 3.4. Detector

1. An either/or proposition for either flow-proportional or scintillation detector.
2. PHA settings: Each detector has a common base line but each has a different window width. These are pre-set manually but the proper window pot is automatically engaged for the particular detector called for in the program. Also, the equipment has a sine-function preamplifier so that as the goniometer is changed to a different element, the counting pulse-band is set automatically in the window.
3. High voltage: Two high voltage supplies set manually, one for each detector.

## 3.5. Crystal

1. Analyzing crystals: Since the reflecting range efficiencies of each crystal is limited, it is advantageous to have a variety of crystals. For this particular project: KAP, AdP, PET and LiF crystals were used.

## 3.6. Collimator

1. An either/or proposition depending on resolution required and intensity of secondary radiation.

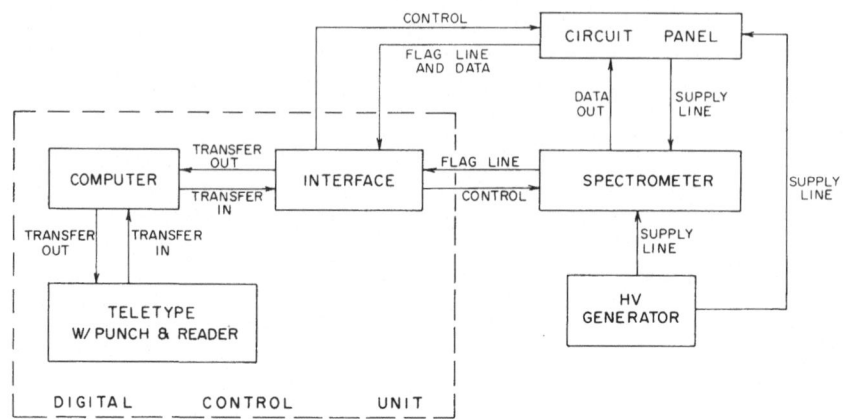

Figure 4. Block diagram of a computer controlled X-ray spectrometer.

## 4. COMPUTER PROGRAM

Figure 4 is a block diagram showing the major components of a computer-controlled X-ray spectrometer system. The large dashed block on the left, termed the Digital Control Unit, represents the computer equipment and the other three blocks are the X-ray equipment. If the diagram was also considered as a flow chart, the normal starting point would be the typewriter or punch reader. The software programs are read in, starting address is set on the switch register, the load address key depressed and then the start key. The individual program containing desired elements, sequence, measuring conditions, and so forth, are then read in on the typewriter or, after a program is established, it could be put on punched tape and read in.

The computer relates these commands to the interface to set the wanted conditions in the spectrometer and circuit panel. The interface is a translator between the computer and peripheral equipment and the major components in the interface are Device Selectors; standard modules made by Digital Equipment Corporation which are altered for the particular function. After a measurement is taken, the data is fed back to the computer.

A flag line or a flag are terms denoting that certain conditions must be met before continuance of the next sequence and usually involve some sort of time delay. For conditions such as changing crystals, collimators or detectors, a flag is set and not removed until a signal is given that the change is complete.

Siemens provided with the unit two software programs. One to control the operating parameters and the other to determine and print-out concentration directly.

The first program provides for the automatic execution of the analysis for different elements in different samples with varying operating conditions. The operating conditions were described previously (Fig. 2) but are relisted here: $2\theta$ setting of goniometer, collimator (2 choices), analyzing crystal (8 choices—1st and 2nd order reflections of 4 crystals), sample rotation (on–off), sample location (8 samples) and detector (2 choices). Those conditions not controlled automatically, and which must be pre-set before a run, are the kilovoltage (kV) and milliamperage (mA) for the X-ray tube, pulse-height analyzer settings and counting mode.

The program has a sequence field of $256_{10}$ (subscript denotes number system: 10-decimal, 8-octal, 2-binary). The computer memory units are flip-flop devices utilizing the binary system. For simplifications and convenience, the octal number system is used in programing places in locations $1200_8$ to $1577_8$. The addresses of the operating conditions are stored there in the sequence in which the elements are to be measured. Also, depending on the matrix and on the concentration range, different conditions and values can be specified and stored separately. A total of $128_{10}$ storage positions are available for these values and each storage location is addressed by a letter (A–G) and the element (2 digits).

The other program provides the facility to calculate elemental concentrations from the intensity values by determining the slope and background constants. There is also a program for taking background measurements, averaging and subtracting from the peak count before calculating the slope of the curve. The output for this program lists the background count, peak, the other background count, net count, and concentration.

An operator's option, included in the program, is a method to combine preset time with preset counts. Figure 5 shows three samples with varied intensities. Sample "a" had sufficient concentration to give us the desired quantity of counts in one time period, sample "b" took three time periods and sample "c" never got to the preset count-limit. The print-out for b and c will be the average one (one basic time). Where to set these limits depends on the individual analysis and the statistics required. Although this facility is provided and is extremely useful, we usually pick compatible concentrations for each inlist program. These limits are set in the switch register and deposited in certain locations or they can be set in by means of the ODT program.

Another feature provided in the software are options to determine the method of calculation and whether to repeat the sequence or not, Fig. 6. In the first method of calculation, the standards are always in position, their concentrations stored and the unknowns are compared directly to them. In the second method, the slope and background constants have been stored previously and all samples are treated as unknowns.

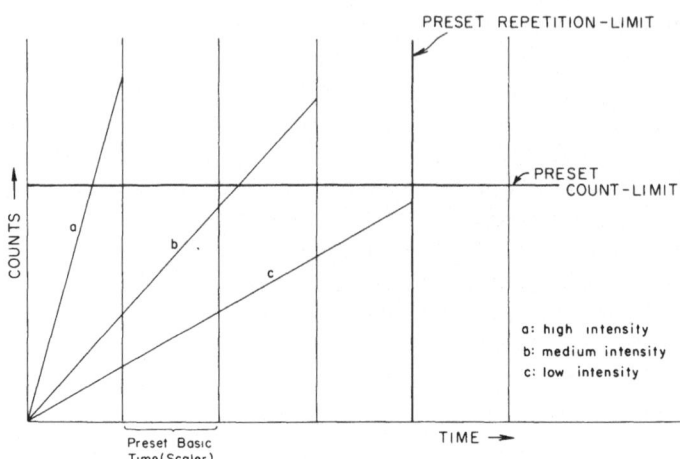

Figure 5. Repetition count-limit schematic.

Figures 7 and 8 show examples of the input–output listings. In the first example, the unknowns are being compared directly to the standards while in the second example, the composition of the standards have been stored previously.

## 5. SAMPLE PREPARATION—GEOLOGICAL MATERIALS

For bulk samples the sample preparation technique has been previously described ([1,2]). Briefly, the method consists of dual grinding a sample as shown in Fig. 9 and making a pressed powder pellet. In making the pellet, a gram of sample is placed in a special die, a small amount of Bakelite is poured around and on top of the powder and then compressed at 30,000 psi.

| SWITCH | | | | MODE |
|---|---|---|---|---|
| 0 | 1 | 2 | 4 | |
| UP | | | | Pos. 0 and Pos. 1 contain standards |
| DOWN | | | | All samples are unknowns |
| | UP | | | Calculate slope and background constants and store in Matrix - Element - Field |
| | DOWN | | | Store concentration of Standards in Matrix - Element - Field |
| | | UP | | Continuous repetition of measuring–sequence |
| | | DOWN | | Stop after completion of measuring–sequence To repeat, depress "CONT." key |

Figure 6. Switch options.

```
                              INPUT  -  LIST

>
1201←1200
POS   0  MATRIX    A ELEMENT   14  (  MC   DEF 52 20:109 060 ↑  %=52 640 ↑ /
POS   1  MATRIX.   A ELEMENT   14  (  MC   52 20:109.060 %=72.650 ↑ /
POS   2  MATRIX    A ELEMENT   14  (  MC   52 20:109 060 /
POS   3  MATRIX    A ELEMENT   14  (  MC   52 20:109.060 /
POS   4  MATRIX    A ELEMENT·   14  (  MC   52 20:109 060 /
POS   5  MATRIX    A ELEMENT   14  (  MC   52 20:109.060 /
POS   6  MATRIX    A ELEMENT   14  (  MC   52 20:109.060 /
POS   7  MATRIX·   A ELEMENT   14  (  MC   52 20:109 060

THE CONDITIONS LISTED SHOW THAT THERE ARE EIGHT SAMPLES AND ARE IN
COMPUTER LOCATIONS 1200 TO 1207   COMPUTER WANTED TO START AT
LOCATION 1201 BUT WAS CHANGED TO 1200.

THE MATRIX WAS DEFINED AS A, ELEMENT IS SILICA

MEASURING CONDITIONS  52 THE FIVE DEFINES 0 15° COLLIMATOR, FLOW
PROPORTIONAL COUNTER AND SAMPLE ROTATION   THE TWO DEFINES CRYSTAL
POSITION 2  FIRST ORDER (PET)


                              OUTPUT  -  LIST

<
1200 TO 1207
A14       A14       A14       A14       A14       A14       A14       A14
142264    223216    207742    201442    204628    210952    208802    206088
52 640*   72 650*   68 642    66 293    67.657    68.881    69.036    69 070

FIRST LINE IS COUNTS PER TIME INTERVAL
SECOND LINE IS PERCENT COMPOSITION OF SILICA   "*" DEFINES
STANDARD

ALL INFORMATION THAT MUST BE TYPED BY THE OPERATOR IS UNDERLINED
```

Figure 7.  Example of input–output list.

No binder is used in the sample itself. For small samples, such as those scraped from the surface of a rock, a fusion method ([3]) is preferable.

## 6. COMMENTS

Whether the choice to have a computer control a piece of analytical equipment or not has to be based on the user's knowledge, experience and whether or not he thinks it is worth it.

```
                              INPUT  -  LIST

>
1210↑
POS  0 MATRIX   B ELEMENT   19  (  MC   DEF 52 20:50.642 ↑ %=0.640 ↑ /
POS  1 MATRIX   B ELEMENT   19  (  MC   52 20:50 642 %=5 430 I

>
1212↑
POS  0 MATRIX   B ELEMENT   20  (  MC   DEF 52 20:45.151 ↑ %=1.360 ↑ /
POS  1 MATRIX   B ELEMENT   20  (  MC   52 20:45.151 %=10.940 I

>
1214↑
POS  0 MATRIX   C ELEMENT   26  (  MC   DEF 70 20:57.524 ↑ %=1 960 ↑ /
POS  1 MATRIX   C ELEMENT   26  (  MC   70 20:57.524 %=11 100 ↑

>
1216↑
POS  0 MATRIX   B ELEMENT   19  (  MC   52 20:50 642 )
                            20  (  MC   52 20:45 151 /
POS  1 MATRIX   B ELEMENT   19  (  MC   52 20:50.642 )
                            20  (  MC   52 20:45.151 /
POS  2 MATRIX   B ELEMENT   19  (  MC·  52 20:50.642 )
                            20  (  MC   52 20:45.151 /
POS  3 MATRIX   C ELEMENT   26  (  MC·  70 20:57.524 /
POS. 4 MATRIX   C ELEMENT·  26  (  MC.  70 20:57.524
```

Figure 8.  Example of input list.

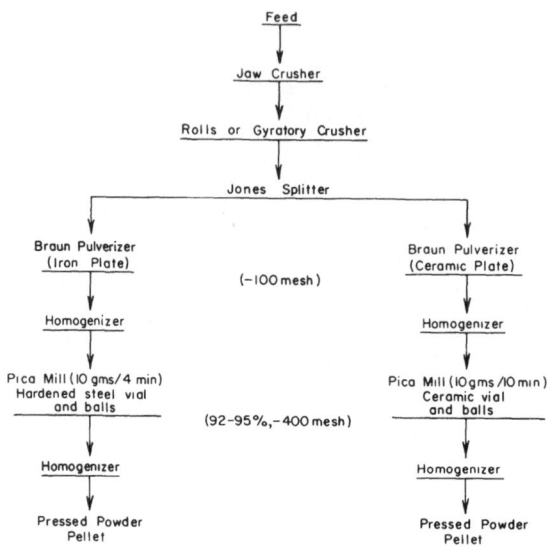

Figure 9. Sample preparation flow diagram.

One of our primary reasons in purchasing the computer type is that we would also have a general-purpose digital computer for other purposes. Of course, it is an either/or proposition, but this equipment was bought for one of the National Aeronautic and Space Administration's Remote Sensing Projects ([4]) and, at the time, a large number of rock samples were collected and each was analyzed for the major constituents. Therefore, an automatic unit was essential. Since that time, the load of samples has decreased and the computer has been used for some project-related data processing.

## REFERENCES

1. A. Volborth, Total Instrumental Analysis of Rocks, Nevada Bureau of Mines Report 6 (1963).
2. P. A. Weyler, Silicate Analysis by X-ray Spectrography, Nevada Bureau of Mines Report 13, Part B (1966).
3. H. J. Rose, I. Adler, and F. J. Flanagan, *Appl. Spectry.* **17**, 81–85 (1963).
4. P. A. Weyler, Computer Controlled X-ray Spectrometric Analysis of Geological Materials, NASA Technical Letter #11, Mackay School of Mines, University of Nevada (1968).

# V. COMPUTER INTERFACE AND DIGITAL SWEEP FOR AN NMR SPECTROMETER

Richard C. Hewitt

*Bell Telephone Laboratories, Incorporated*
*Murray Hill, New Jersey*

The powerful technique of using a small digital computer to enhance the signal to noise ratios of weak NMR signals is well known. In short, the NMR spectrometer–computer interface consists of an integrating type analog to digital converter to digitize the spectrometers output, a digital to analog converter to display the collected spectra, and a digital sweep system to control the frequency or field of the NMR spectrometer. The latter also serves as the timing element used to read the analog to digital converter and to advance the channel of a multichannel analyzer program.

Under program control the digital sweep control circuit selects the frequency of a Hewlett–Packard Frequency Synthesizer and is capable of producing a highly linear reproducible sweep with accurate frequency markers.

Initially the system was implemented using a Digital Equipment Corporation PDP-8 computer having a 4K work memory. Later a 128K word magnetic drum memory and a dataphone interface were added. The former increases the storage capabilities of the system and the latter provides the powerful computational facilities of a large processing system, while an experiment is in progress.

Details of the system will be presented.

## 1. INTRODUCTION

The powerful technique of using a small digital computer to enhance the signal-to-noise ratio of weak NMR signals is well known. However, NMR spectrometer-computer interfaces differ widely depending on the NMR spectrometer and design considerations. The interface to be described utilized the remote digital programing feature of a frequency synthesizer to produce a highly linear and reproducible sweep with accurate frequency markers. Obviously, a purely digital system precludes a strictly continuous sweep and actually the sweep consists of up to 100,000 discrete frequencies (depending on range setting). The digital sweep control circuit is capable of changing the frequency of the frequency synthesizer every $10\,\mu$sec. It was built from standard computer-type logic modules, and thus is very flexible in respect to any modifications in logic that may be required for a specific purpose. This mode of construction makes it also easy to control the sweep functions from an on-line computer if so desired. In the present version, the starting frequency of the sweep can be set and the sweep started and stopped on computer command. Conversely, the sweep unit controls computer functions through the program interrupt. It also provides the gating for an integrating type analog to digital converter used to digitize the spectrometer's output.

Other sweep functions such as setting the sweep time and width, are at present by manual switches. However, standard logic voltage levels are used throughout, and it is thus a straightforward matter to replace the switches by corresponding flip-flop registers, to obtain complete control through input–output commands of the computer.

## 2. SYSTEM IMPLEMENTATION

In computer averaging of NMR spectra it is necessary to keep the line positions with respect to the start of the sweep constant for many sweeps through the spectrum. If the line positions change between successive sweeps, the lines in the averaged spectra will be broadened. To avoid this line broadening an external field lock was added to the Varian HR-60 NMR spectrometer. Basically, it is a sideband oscillator ([1]), locking on a fluorine resonance to follow any changes in the magnetic field. Negative feedback is also given to the superstabilizer to compensate for changes in the magnetic field. The ratio 17 : 16 of proton resonance to fluorine resonance is utilized to derive a 60 MHz reference for protons. The addition of an audio frequency (5 kHz) obtained from the frequency synthesizer provides the radio frequency for proton resonance. Changing the audio frequency will produce frequency sweep. Similarly, changing the audio frequency sideband (10 kHz) of the external fluorine field lock will generate a field sweep.

A PDP-8 general purpose digital computer ([2]) is used for the time-averaging of the NMR signal. This specific computer has a 12-bit word length, 4096 words of random access core memory and a memory cycle time of $1.5 \mu$sec.

The output voltage of the spectrometer is digitized with a Vidar ([3]) voltage to frequency converter, the output of which is counted by a gated binary scaler and read into the accumulator of the computer on application of a transfer pulse. The pulse is generated by the digital sweep control circuit, and it also resets the binary scaler of the analog to digital converter. A 1000 channel analyzer program adds the contents of the accumulator to the sum of the previous counts representing that channel. The channel address is then advanced for the next count and the cycle is repeated for each of the 1000 channels. Double precision (24 bits) is used to store the count per channel.

Half of the computer's memory is used to store the spectra, and the other half contains the control program. The collected spectrum is displayed on a cathode-ray tube. Certain teletype commands have been included in the multichannel analyzer program. Activation of any key is acknowledged only after the sweep stops after completion of the current sweep. The starting frequency can be changed by typing in the new frequency. Other teletype commands include: start sweep, clear memory, adjust the display amplitude, print the number of scans collected, and an option to stop the scanning after a predetermined number of sweeps.

## 3. DETAILED CIRCUIT DESCRIPTION

The instrument as built produces triangular sweeps with widths ranging from 100–100,000 Hz in 1, 2, 5, 10 ... steps. Forward and reverse sweep times can be set independently from 1–1000 sec, in the same steps mentioned above. In addition there is a fast reverse setting of 0.1 sec. The frequency linearity and reproducibility is $\pm 0.01$ Hz over a range of 100 kHz, and the linearity with respect to time is better than one part in $10^9$.

A number of features which were thought to be convenient for our application were provided, but these could be easily deleted, modified, or added to as desired. A STOP push button will stop the sweep at any time. On pushing the START switch the periodic sweep will resume where it left off. The RESET SWEEP is a simple reset of the sweep to its starting frequency. The SET FIRST FREQUENCY switch, when pushed will reset the sweep without resetting the frequency so as to make the frequency at the moment of activation the start frequency of the sweep. The SET LAST FREQUENCY will similarly set the final frequency of the sweep. The effect of these switches is to shift the sweep's range to lower or higher frequency, respectively, and they are used for searching and for centering a display. In the MANUAL position of the SWEEP MODE switch, the sweep is under manual control, the forward or reverse sweep continuing as long as the FREQUENCY SELECTOR switch is held in the corresponding position. A seven-digit numerical display, activated by the same voltage that sets the synthesizer frequency, indicates the output frequency at any time. A digital to analog converter gives an output voltage proportional to the frequency, and is used for driving the x-deflection of an oscilloscope display.

The frequency synthesizer used was the Hewlett–Packard Model 5102A [4]. It was modified by the manufacturer (H75 option) to provide a means for obtaining a phase-coherent output while switching. That is, the phase of the new output will be the same as the phase of the old one when switching occurs. This modification provides a means to align the phases of the ten bus frequencies in the synthesizer to have the same phases every 10 $\mu$sec. The desired frequency is obtained by selecting one of these 10 bus frequencies for each decade of the synthesizer [5]. Since only the bus frequencies are switched, and their phases are the same when switching occurs, the output is continuous, that is phase-coherent. This modification required that the switching in the control circuit be synchronized with the bus frequency phases in the synthesizer. A 100 nsec clock pulse with a pulse repetition rate of 100 kHz was derived from the synthesizer crystal oscillator for this purpose.

The major part of the control circuitry was built from "Flip Chip" [2] logic modules. One special module, to be described below, was homemade. Figure 1 represents two out of seven identical decades of the control

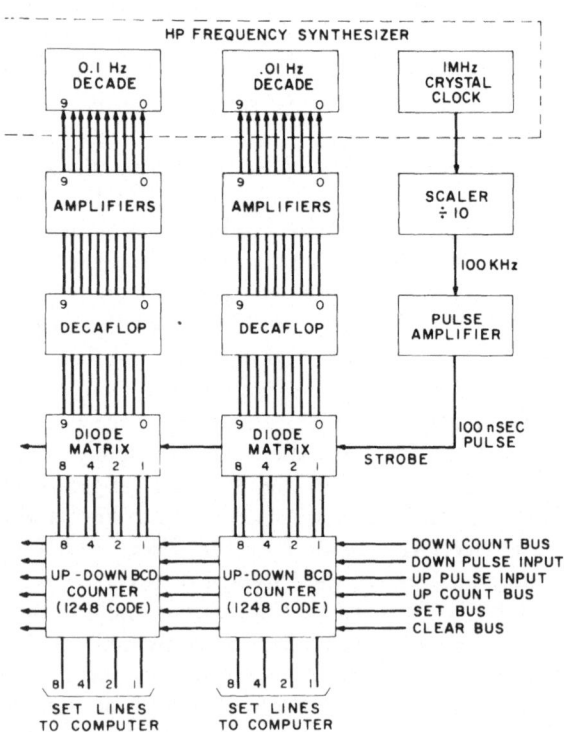

Figure 1. Block diagram of control circuit for the frequency synthesizer. Two of the seven decades are shown.

circuit for the frequency synthesizer. At the bottom of Fig. 1 is a BCD up–down counter (7 decades) which can be made to count up or down by applying the assertion level to the corresponding count bus and pulsing the respective input. The sweep rate is proportional to the repetition rate of the input pulses. The sweep width is controlled by reversing the direction of the counter after a predetermined number of pulses. Doing this repetitively allows one to obtain a recurrent sweep. A fast retrace is obtained by increasing the pulse repetition rate in the down direction.

The BCD counter can be reset to the zero state, or set to any BCD number, by applying the BCD information to the set lines and giving a set pulse. In this mode a digital computer can control the contents of the BCD counter or buffer. The BCD information of the counter is converted to decimal 10 line by a diode matrix. The diode matrix is gated off until the 100 $n$sec clock pulse strobes the contents of the up–down counter into a decimal register (consisting of a chain of ten-state units we call "decaflops").

Figure 2 shows the configuration of the up–down BCD counter (1248 code) using the standard DEC notation ([6]). The positive and negative assertion levels are represented by the open and solid diamonds, respectively. The open arrows represent a positive-going pulse. The "0" and "1" states are represented by $-3$ and 0 V, respectively, at the 0 output of a flip-flop. If the assertion level as shown in Fig. 2 is supplied to the up-count enable bus, the carry circuits in the decade are enabled. By applying positive-going pulses to the up pulse input, the counter will count with increasing BCD number. Down counting is accomplished in a like manner. By removing the assertion levels from both the up and down count enable buses, all the carry circuits are disabled. The counter can then be used as a buffer register for computer control of the synthesizer frequency. A 100 $n$sec positive-going pulse applied to the clear bus will reset the flip-flops to the 0 state. Applying a positive assertion level to the set line of a flip-flop will cause it to be set to the 1 state when a 100 $n$sec positive-going pulse is applied to the set bus. This allows BCD information to be transferred from a computer to the register.

Figure 3 shows a decaflop which is a ten-state circuit. It is a generalization of a flip-flop and consists of ten transistors. The output of the decaflop is amplified to obtain the levels required to control the frequency synthesizer. The amplifier also serves to reduce the capacitive loading on the decaflop. Seven such circuits make up the register which contains the frequency information in the ten-line code required by the frequency synthesizer.

The bottom part of Fig. 3 shows the diode matrix which converts the 1248 BCD code of the counter to decimal ten line code. The diode matrix is disabled, that is, its output is held at 0 V, until a 100 $n$sec clock pulse arrives. The clock pulse is a negative pulse ( $-3$ V), thus allowing one of the

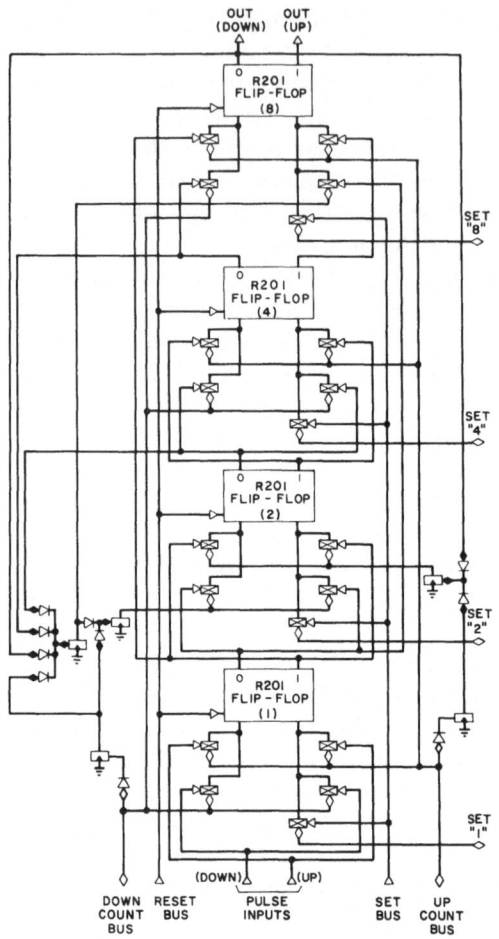

Figure 2. Schematic diagram of a BCD decade used in the up–down counter.

ten outputs to go to −3 V. This causes 1 mA of base current to flow in the corresponding Q2 transistor, and causes the collector circuit to pass its saturation current. The collector of the same Q2 is now at 0 V and keeps the other nine transistors from conducting by the diode feedback network shown above the Q2's in Fig. 3. The outputs of the nonconducting Q2's are clamped to −3 V by the diodes shown to the left above the Q2's. That is, the base is cut off by a 0.5 V positive voltage with respect to the emitter, due to the voltage drop across the diodes D2. The output taken from the collectors of the Q2's is converted by the Zener diode D1 and amplifier by Q1. The levels on the collector of Q1 are 1.5 and −12.6 V. The "on" state is represented by the −12.6 V. An output taken from this point is used to select the bus frequencies, that is, the digits in the frequency synthesizer.

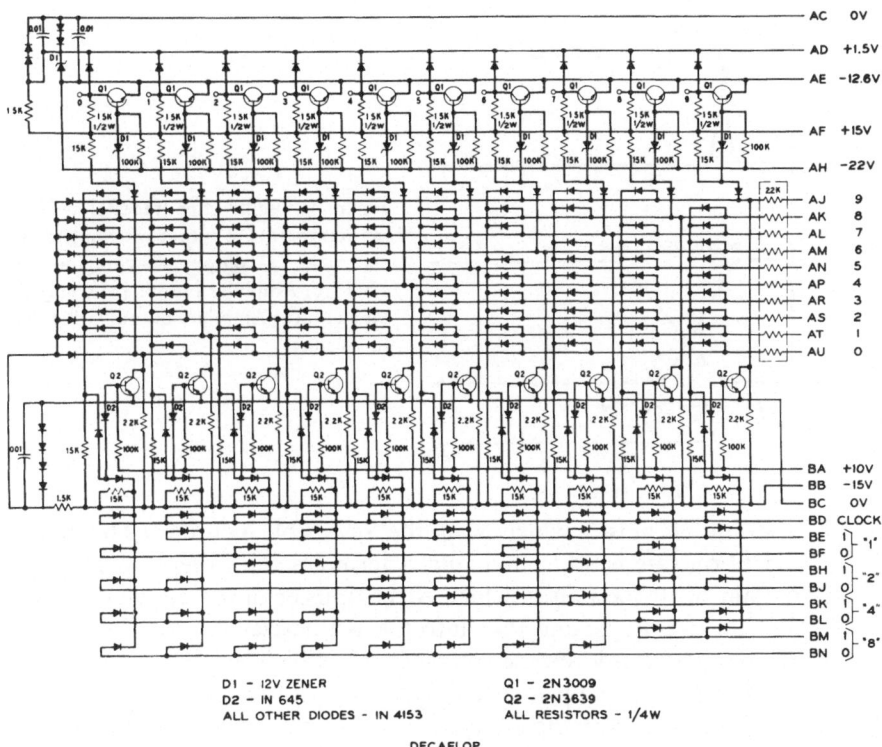

D1 – 12V ZENER                    Q1 – 2N3009
D2 – 1N 645                       Q2 – 2N3639
ALL OTHER DIODES – 1N 4153        ALL RESISTORS – 1/4W

DECAFLOP

Figure 3. Diagram of the BCD to decimal diode matrix, Decaflop and output amplifier.

The total switching time for the decaflop is 50 nsec. Figure 4a and b are waveforms obtained at the collectors of two of the Q2. The (a) waveform is the turning on of a Q2 and the (b) is the turning off of another. Figure 4c and d are the outputs of the Q1 amplifiers associated with the Q2 above. The (d) waveform shows the switching to the "on" state and the (c) to the "off" state. The total time for the amplifiers to quiet down is 300 nsec. Switching will occur in the modified synthesizer when the voltage goes through 0 V. Therefore, the total switching time is approximately 200 nsec. A typical output from the synthesizer is shown in the upper trace in Fig. 5. The frequency is being switched in 10 kHz increments from 9866.67–99,866.67 Hz and back to the first frequency repeatedly. The lower trace in Fig. 5 shows the switching voltage applied to the 0 digit of the 10 kHz decade in the synthesizer. It can be seen from this photograph that the output is continuous (or phase coherent) during switching.

The sweep system described below has as its function the provision of the succession of pulses and levels which takes the reversible counter through the desired sweep program. The voltages supplied are:

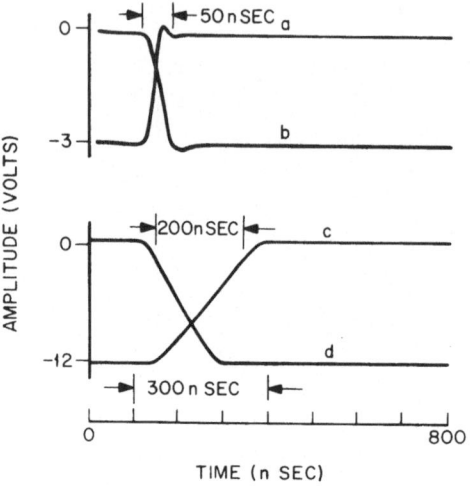

Figure 4. Oscilloscope traces of switching action: (a) Q2 output switching to on state; (b) Q2 output switching to off state; (c) Amplifier output switching to off state; (d) Amplifier output switching to on state.

1. A train of pulses which step the reversible counter. The pulse repetition rate is adjustable and determines the sweep time.
2. A train of strobe pulses, which effect transfer of the counter setting into the decaflops, and thus into the synthesizer.
3. A pair of forward–reverse voltage levels, which switch the reversible counter from the up to the down count status and vice versa, as required for a periodic sweep.

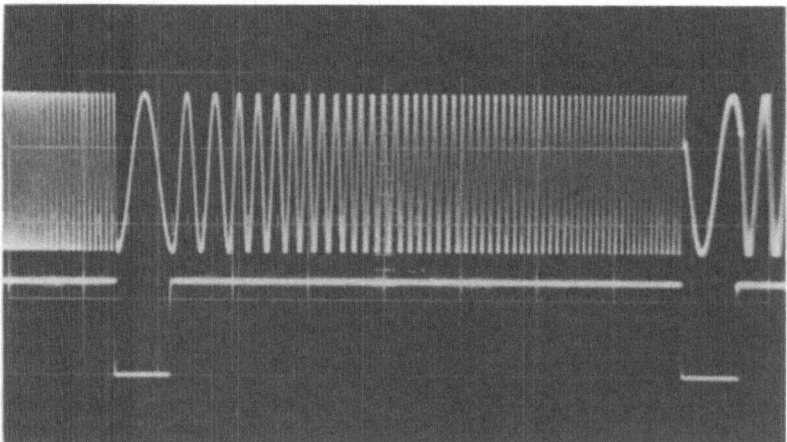

Figure 5. Upper trace: Output of the Frequency Synthesizer while the frequency is switched repeatedly in 10 kHz steps from 9,866.67 Hz to 99,866.67 Hz, and back to the first frequency. Lower trace: Switching voltage applied to the "0" line of the 10 kHz synthesizer stage.

4. A level output which effects synchronization of a time averaging computer with the sweep. This output also operates a gate in the analog to digital converter used in conjunction with the time averaging computer.

A detailed description follows. A block diagram of the digital sweep circuit is given in Fig. 6. A 100-kHz clock frequency is derived from the 1-MHz crystal oscillator in the frequency synthesizer. This is done by dividing the 1-MHz signal by 10 with the CLOCK SCALER which produces a square wave output at 100 kHz. The negative-going transitions provide the CLOCK pulses and the positive-going ones, provide the CLOCK pulses. The latter are thus offset by 5 $\mu$sec from the former. An output taken from the CLOCK SCALER passes through GATE 1 which can be disabled by the computer and cause the sweep to stop. When enabled, the clock frequency is applied to the inputs of both the FORWARD and

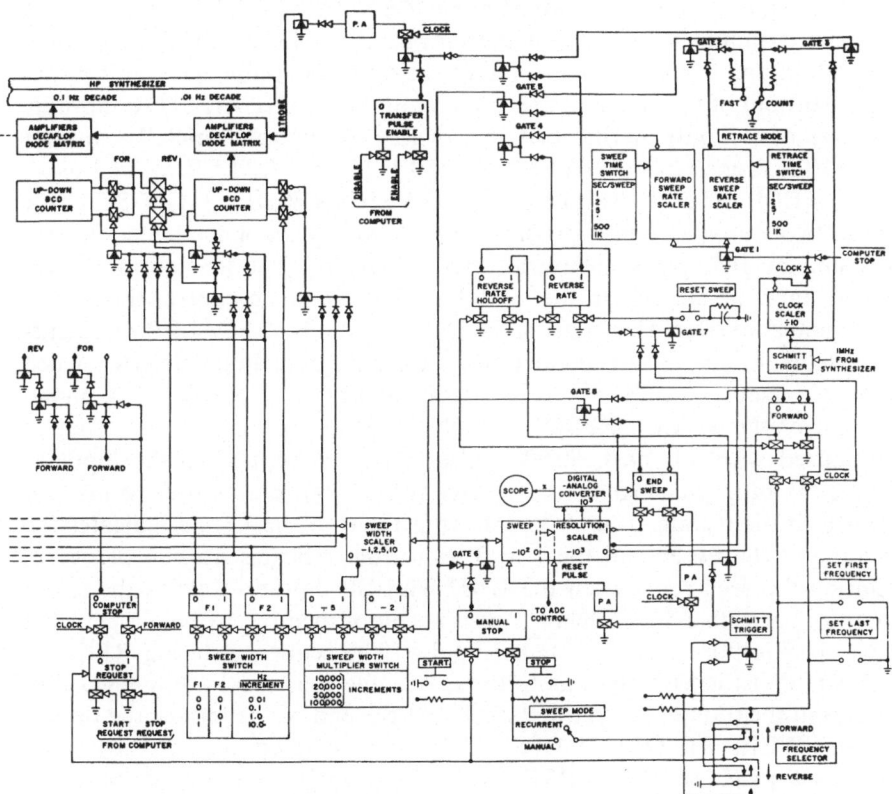

Figure 6. Block diagram of the digital sweep circuit.

REVERSE SWEEP RATE SCALERS. The two scalers are identical and divide the input frequency by a factor between 1 and 1000 in $1:2:5:10\ldots$ steps, depending on the setting of the SWEEP TIME SWITCH and the RETRACE TIME SWITCH, respectively. The output of the REVERSE SWEEP RATE SCALER goes to GATE 2 which is enabled or disabled by the RETRACE MODE SWITCH. The output of GATE 2 is OR-ed with the output of GATE 3 which has 1 MHz applied to the input and also enabled by the RETRACE MODE SWITCH. With the RETRACE MODE SWITCH in the COUNT position, the output of the REVERSE SWEEP RATE SCALER appears on the output of the gates. In the FAST REVERSE position, the 1 MHz is present at the output of GATES 2 and 3. This output, and the output of the FORWARD SWEEP RATE SCALER go to GATES 4 and 5, which selects one of them according to the status of the REVERSE RATE flip flop. If the latter is in the 1 state, the reverse rate is selected, while in the 0 state the forward rate is selected. This output goes to GATE 6, which is enabled if the MANUAL STOP flip-flop is in the 0 state. It should be noted that the output of GATE 6 is a positive going transition and corresponds in time to the CLOCK pulse. To avoid pulses in the output of the gate due to switching of the MANUAL STOP flip-flop, the latter is set only on the positive transition of the gate input. The output of GATE 6 goes to the SWEEP RESOLUTION SCALER as well as to the SWEEP WIDTH SCALER. The SWEEP RESOLUTION SCALER gives two outputs: one a division by $10^2$, and a second after additional division by $10^3$, giving an overall division by $10^5$. The first output serves to advance the channel in the digital computer and to control the analog to digital converter. The status of the DIVIDE BY 1000 SCALER represents the channel number. It is connected to a digital to analog converter which supplies the drive to the $x$-axis of a scope and an $x$–$y$ recorder. The digital output of the SWEEP RESOLUTION SCALER is symmetrical. That is, initialized with all zeros, 50,000 pulses must be applied to the input to make the output change to the 1 state and 50,000 more pulses to make it return to the 0 state. The switching from the 0 to the 1 state indicates the middle of the sweep and the 1 to 0 transition indicates the end of sweep. During the recurrent sweep the END SWEEP flip-flop follows the scaler output, but with a 5 $\mu$sec delay. This is accomplished by strobing the output into the flip-flop with the $\overline{\text{CLOCK}}$ pulse. The sequence during RECURRENT SWEEP MODE is as follows. Assume that the FORWARD flip-flop is in the 0 state, i.e., retrace status. At the end of the retrace the output of the SWEEP RESOLUTION SCALER switches from the 1 to the 0 state, and as a result the REVERSE RATE flip-flop is set to the 0 state too. In this state GATE 4 applies the output of the FORWARD SWEEP RATE SCALER to GATE 6. Five $\mu$sec later the END SWEEP flip-flop is set to the 1 state. This sets the REVERSE RATE HOLDOFF flip-flop to the 0 state and toggles the FORWARD flip-flop to the 1 state. The FORWARD flip-flop controls the direction of counting in the UP–DOWN COUNTER which holds the digits for the frequency synthesizer. Sweep is now in the forward direction. At midsweep the output of the SWEEP RESOLUTION SCALER switches to 1, and as a result the END SWEEP flip-flop switches to 0. Forward sweep continues until, at the end of the forward sweep, the output of the SWEEP RESOLUTION SCALER switches again to the 0 state. This, along with the 0 state of the REVERSE RATE HOLDOFF flip-flop, enables GATE 7 which will set the REVERSE RATE flip-flop to the 1 state when the FORWARD flip-flop switches to the 0 state. With the 5 $\mu$sec delay previously mentioned, the END SWEEP flip-flop will switch to the 1 state. This will toggle the FORWARD flip-flop to the 0 state, thus setting the REVERSE RATE flip-flop to the 1 state. With the reverse sweep rate now selected and the direction of the UP–DOWN COUNTERS reversed, the sweep will retrace. When the SWEEP RESOLUTION SCALER output switches to the 0 state again, the cycle is repeated. An input for the SWEEP WIDTH SCALER is taken from the same point as the input for the SWEEP RESOLUTION SCALER. Exactly 100,000 pulses must appear at this point during each direction of sweep. The SWEEP WIDTH SCALER divides this by a factor of one, two, five, or ten depending on the states of the DIVIDE BY 5 and DIVIDE BY 2 flip-flops. These flip-flops are set at the middle of the forward sweep by the SWEEP WIDTH MULTIPLIER switch. The set pulse is obtained by AND-ing the 0 and 1 outputs of the END SWEEP and FORWARD flip-flops, respectively. The same pulse is used to set the F1 and F2 flip-flops from the setting of the SWEEP WIDTH switch. These flip-flops determine onto which of the first four decades of the UP–DOWN COUNTER the output of the SWEEP WIDTH SCALER is applied. The described arrangement insures that the sweep width switching is executed at midsweep and thus that the center frequency stays fixed and independent of the sweep width. The SWEEP WIDTH switch selects steps of 0.01, 0.1, 1, or 10 Hz and the SWEEP WIDTH MULTIPLIER switch selects 10,000, 20,000, 50,000, or 100,000 steps in a sweep. The product of these two switch settings gives the sweep width, which can range from 100–100,000 Hz. It should be noted that the input to the UP–DOWN COUNTER is coincident with the CLOCK pulse. The $\overline{\text{CLOCK}}$ pulse, 5 $\mu$sec later, generates the STROBE pulse which transfers the contents of the UP–DOWN COUNTER into the DECA-FLOPS. A pulse amplifier converts the $\overline{\text{CLOCK}}$ pulse into a 100 $n$sec pulse, which is inverted to obtain the negative-going STROBE pulse. The STROBE pulse is inhibited when the TRANSFER PULSE ENABLE flip-flop is in the 0 state. This flip-flop is set by the computer and is used to disable the STROBE pulse during the setting of the UP–DOWN COUN-TER under computer control. The STROBE pulse is also disabled during the retrace when the retrace mode switch is set to FAST RETRACE MODE. This is necessary to avoid erroneous setting of the frequency synthesizer, because in this mode a STROBE pulse can occur while a carry propagates in the up–down counter.

A stop request can be made by the computer by setting the STOP REQUEST flip-flop to the 1 state. This sets the COMPUTER STOP flip-flop to the 1 state, when the FORWARD flip-flop switches to the 1 state. This in turn disables GATE 1. Thus, the sweep is stopped at the beginning of sweep and at the first frequency. In this mode the computer has complete control of the frequency synthesizer, and flexible frequency programming can be done.

In one application the computer sets the first frequency of sweep and returns control to the sweep unit by giving a START REQUEST pulse. This pulse causes the STOP REQUEST flip-flop to be set to the 0 state when the $\overline{\text{CLOCK}}$ transition occurs. Since the FORWARD flip-flop is switched by the same pulse, the sweep resumes at the same point it was stopped. However, the first frequency is that set by the computer.

The frequency can also be changed by the FREQUENCY SELECTOR switch. This is a three position switch shown in the rest position. The FORWARD and REVERSE position only differ in that the FORWARD position sets the FORWARD flip-flop to the 1 state and the REVERSE to the 0 state. This is done on the $\overline{\text{CLOCK}}$ pulse, so that switching direction of the UP–DOWN COUNTERS while a carry propagates is avoided. When the FREQUENCY SELECTOR switch is activated, the $\overline{\text{CLOCK}}$ pulse, which sets the END SWEEP flip-flop, is inhibited. The END SWEEP flip-flop is set to the 0 state and the REVERSE RATE HOLDOFF flip-flop is switched to the 1 state. The latter holds the REVERSE RATE flip-flop in the 0 state, thus selecting the forward sweep rate in both directions. The MANUAL STOP flip-flop is set to the 0 state and the STOP REQUEST flip-flop is set to the 0 state. The sweep will start and continue in the same direction until the switch is released. Releasing the switch from the FOR-WARD position sets the last frequency of sweep, and from the REVERSE position set the first frequency of sweep. Both cases reset the SWEEP RESOLUTION scaler to zero and causes the $\overline{\text{CLOCK}}$ pulse to set the END SWEEP flip-flop to the 1 state. This sets the REVERSE RATE HOLDOFF flip-flop to the 0 state. With the switch being released from the FORWARD position, the FORWARD flip-flop is toggled to the 0 state, which sets the REVERSE RATE flip-flop to the 1 state. Thus, the unit is at the beginning of retrace. With the switch released from the REVERSE position, the FORWARD flip-flop is toggled to the 1 state and the

REVERSE RATE flip-flop remains in the 0 state. Thus, the unit is at the beginning of sweep. If the SWEEP MODE switch is in the MANUAL position, the MANUAL STOP flip-flop will be set to the 1 state when the switch is released. The SET LAST FREQUENCY and SET FIRST FREQUENCY switches, which are momentarily depressed, have the same effect as momentarily depressing the FREQUENCY SELECTOR switch in the FORWARD or REVERSE direction, respectively. However, they will not start or stop the sweep. During recurrent sweep they serve as a convenient means of shifting the swept frequency.

The RESET SWEEP switch proves to be convenient when long sweep times are selected and one desires the sweep to start over again. Depressing

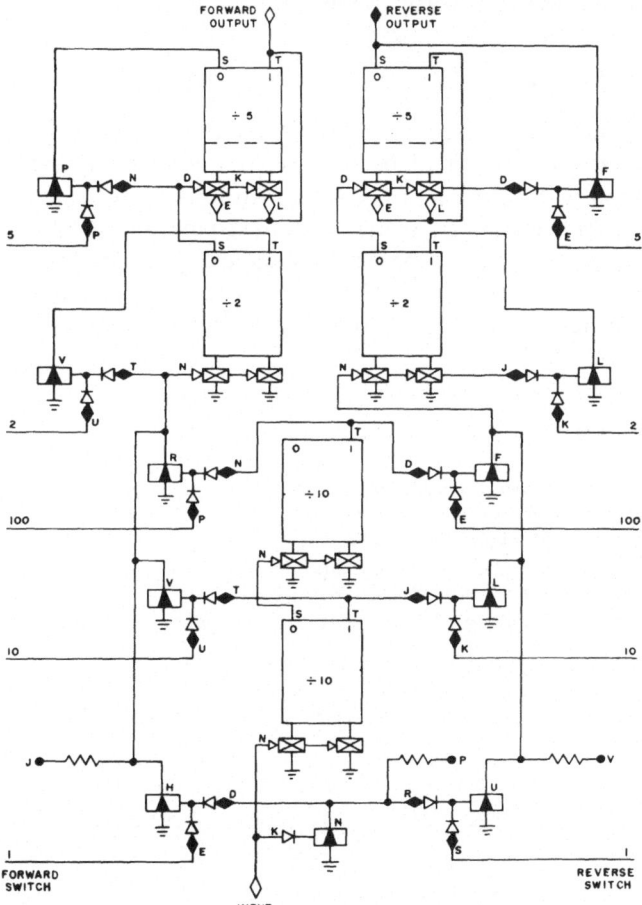

Figure 7. Block diagram of the combined forward sweep rate and reverse sweep rate scaler.

it sets the REVERSE RATE flip-flop to the 1 state. The sweep will continue in the same direction but at the faster reverse sweep rate.

In Fig. 6 and in the previous discussion the FORWARD and REVERSE SWEEP RATE scalers were treated as two discrete units, when in fact they share command elements. Figure 7 is a block wiring diagram of the circuit configuration used. The common elements are two divide by 10 scalers, which are used to divide the input CLOCK frequency by a factor of 10 and 100. The SWEEP TIME and RETRACE TIME switches independently select the input CLOCK frequency divided by 1, 10, or 100 through a set of gates. The selected frequencies are once more divided by a factor of 1, 2, 5, or 10 depending on the switch setting, in order to obtain the desired forward sweep rate and reverse sweep rate.

## 4. CONCLUSION

The circuit as described has been in continuous operation for more than a year. It has proved to be very reliable and highly satisfactory in regard to stability, linearity and flexibility. Although it may seem that these advantages have been gained at the cost of extreme complication, it should be realized that the system is based on standard, well developed computer techniques. It is our experience that the use of proven commercial logic modules is in many ways less demanding than the design of high-quality analog circuits.

## REFERENCES

1. W. A. Anderson, *Rev. Sci. Instr.* **33**, 1160 (1962).
2. Digital Equipment Corporation, Maynard, Mass.
3. Vidar Corporation, Mountain View, Calif.
4. Hewlett–Packard Company, Palo Alto, Calif.
5. V. E. Van Duzer, *Hewlett–Packard J.* **15**, No. 9 (1964).
6. "The Digital Logic Handbook," Digital Equipment Corporation, Maynard, Mass. (1966) pp. 2–32.

# VI. APPLICATION OF THE INFOTRONICS CRS–110/50 COMPUTER INTEGRATOR SYSTEMS FOR ON-LINE GC ANALYSES

J. M. Cotton

*Infotronics Corporation*
*Houston, Texas*

The combination of an automatic digital integrator with a small computer provides a very practical ON-LINE system for automating gas chromatography analyses. Such systems can be started with a small initial investment and expanded as required. The possible independent operation of integrators and computer provide a minimum of down time.

## 1. INTRODUCTION

In 1961, Infotronics Corporation—then a representative for data handling equipment—discovered the need for an electronic digital integrator in gas chromatography. In response to this need, Infotronics engineers designed the CRS-1 integrator. Since then, the company has refined its integration systems (third generation integrators are currently being marketed) and expanded into a variety of on-line and off-line data handling systems for the analytical laboratory. One such system, which teams the accuracy and operating flexibility of the digital integrator with the computational power and speed of a computer, is discussed below. Some uses of magnetic tape playback units in conjunction with a computer are also discussed.

## 2. COMPOSITE INTEGRATOR/COMPUTER SYSTEMS

A typical integrator-computer system is shown in Fig. 1. This particular three-channel configuration was used successfully to automate analyses for a rather elaborate chemical process stream. Functionally, it is very simple. Each integrator performs the timing, detection, and integration of the constituent peaks eluting from its GC or amino acid analyzer. The

# OFF-LINE

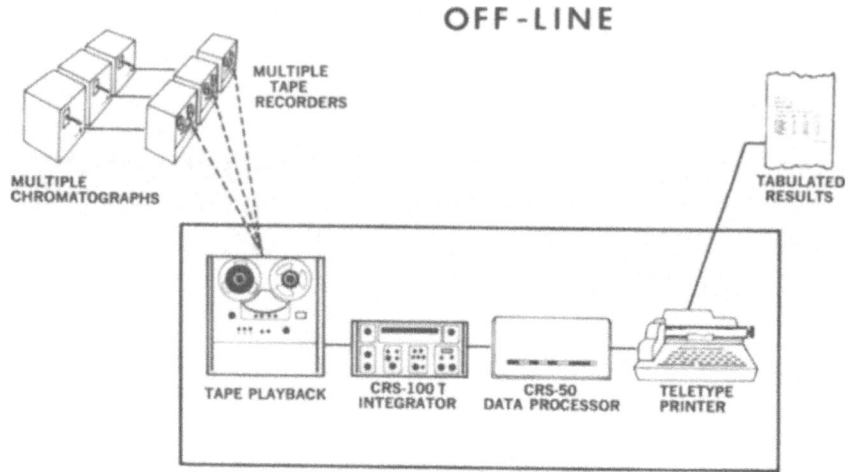

Figure 1.  A typical integrator-computer system.

computer receives the time and area values generated by the integrator and then, when that run is terminated, performs all calculations and comparisons necessary to obtain a complete report. The report is printed out almost immediately and usually includes the sample number, analyzer number, time of run, date of run, and other identification data. Beneath his heading are listed all peaks with names, retention times, peak areas, and analysis results in appropriate engineering units. A sample report, prepared in weight percent values, is shown in Fig. 2.

In the actual operation of the integrator-computer system, the technician injects the sample into the chromatograph and simultaneously presses the "start" button on the interface console. This action causes identification information such as sample type, sample number, analyzer number, etc., to be entered into the computer from a series of thumbwheel switches. Integrator timing is also initiated. As the constituents elute, their peaks are integrated and timed. The result is transmitted in parallel to the printer and to the computer. The software program determines which integrator initiated the transfer, and stores data in an array allocated for that integrator. Figure 3 is a photograph of a typical system center, the CRS-110/50.

When the final peak of interest has been entered into the computer, the technician simply presses a RESET button on the interface console. An *end-of-chromatogram* signal is sent from the reset integrator, notifying the computer that the identification procedures may begin for the particular data. An executive program examines the status of each integrator, and when a *data-completed* switch is found to be on, program control is shifted

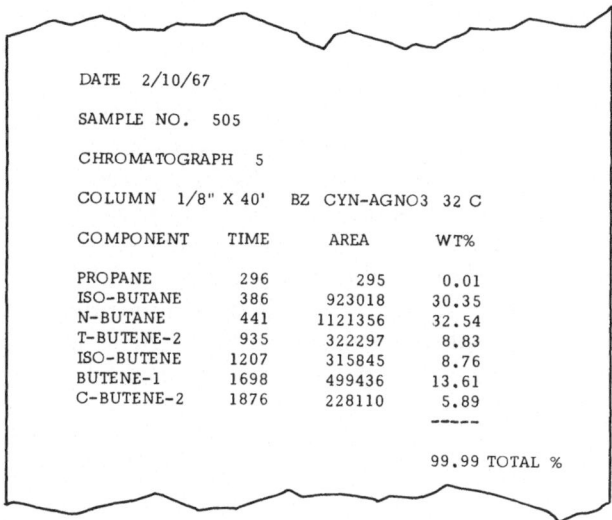

```
DATE   2/10/67

SAMPLE NO.   505

CHROMATOGRAPH   5

COLUMN   1/8" X 40'   BZ  CYN-AGNO3  32 C

COMPONENT      TIME        AREA        WT%

PROPANE         296          295       0.01
ISO-BUTANE      386       923018      30.35
N-BUTANE        441      1121356      32.54
T-BUTENE-2      935       322297       8.83
ISO-BUTENE     1207       315845       8.76
BUTENE-1       1698       499436      13.61
C-BUTENE-2     1876       228110       5.89
                                      -----
                                99.99 TOTAL %
```

Figure 2.  A sample report.

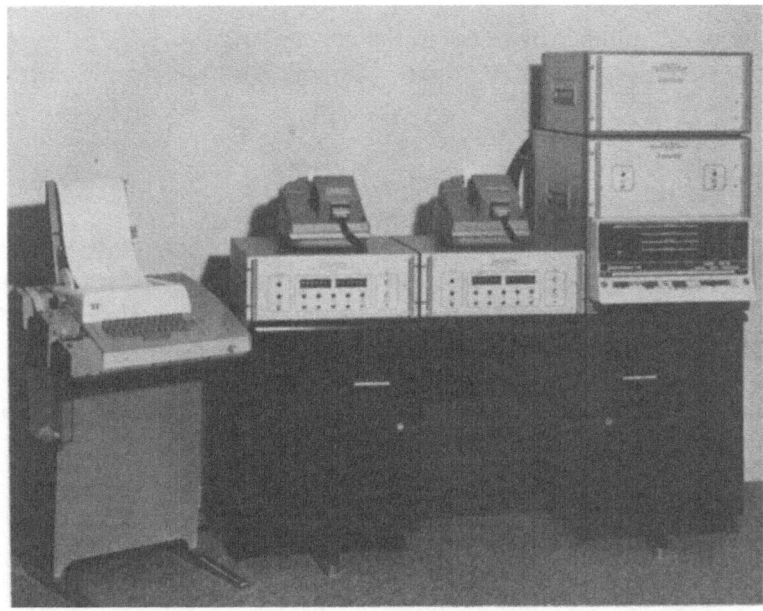

Figure 3.  Photograph of a typical systems center, the CRS-110/50.

to data reduction and identification routines. An interrupt system operates during identification and reporting, so that peak times and areas from other analytical instruments may be entered.

Depending on the method of analysis desired, retention times may or may not be converted to relative values. Normally, relative retention times are used because of the shifts in absolute times. In any case, the temporal position of a given peak to the reference peak is used to identify the former. Only those peaks whose retention times have been programed into the computer will be identified. Extraneous peaks are considered unknown, and reported as such. Once a peak is identified, the proper response factor is applied to obtain its corrected area. Response factors may vary over a wide range, depending on the method of reporting the results. For normalization, equations (1) and (2) are applied. Results are reported in percentages. The area of each peak is proportional to the percentage that the peak comprises in the total sample. The sum of all the areas represents the total sample size. Given an area $A_i$ and response or proportionality factor $W_i$ the size of the total sample may be found by

$$A = \sum_{i=1}^{N} A_i W_i \tag{1}$$

where $N$ is the number of peaks in the chromatogram.

The concentration value of each component in the sample, therefore, becomes

$$\%C = \frac{W_n A_n}{A} \times 100 \tag{2}$$

If an internal standard is used, $A$ is replaced by $A_{is} W_{is}$ in the preceding equation, and 100 is replaced by the percent of internal standard. From the start of the run until the final report is complete, the only manual operation consists of pressing two buttons.

Composite systems offer an excellent answer for small but growing laboratories. Additional integrators can be added as requirements expand without any basic modifications to the existing system.

## 3. MODIFICATIONS TO THE BASIC SYSTEM

The speed of an on-line computer-integrator system may or may not be a desirable feature, depending upon the requirements of the analytical chemist who uses it. In many cases, he may prefer to take advantage of the computer's capability for handling special situations. In this case, an off-line system using magnetic tape may be in order. Such a system, the CRS-40/50, is shown in Fig. 4. Here, the computer and a magnetic tape playback system are mounted in a console—usually remote from the recorder systems at one or more laboratories. Chromatograph runs are recorded, along with digitized identification "headers," in the laboratory. At the conclusion of a run, the tape can be hand carried to the CRS-40/50 console and played back into the computer. Incidentally, this procedure circumvents the cost of installing direct lines—a reasonable trade-off for losses in time.

Because the data is on tape, it can be run over and over again, permitting optimum results for the analyses. Further, any particularly interesting segment of a run can be examined and re-examined. The stored run can be held ready for playback at a much later date. This is an especially useful feature for quality-control laboratories, who may wish to keep a library of tapes.

Direct recording of instrument output signals ordinarily requires exceedingly high-fidelity tape units and effectively increases the signal's exposure to sources of random noise. The answer to this problem lies in preprocessing the chromatogram signals into an FM pulse train impervious to conventional noise. Infotronics magnetic tape systems incorporate the "front-end" of an integrator system—a voltage-to-frequency converter.

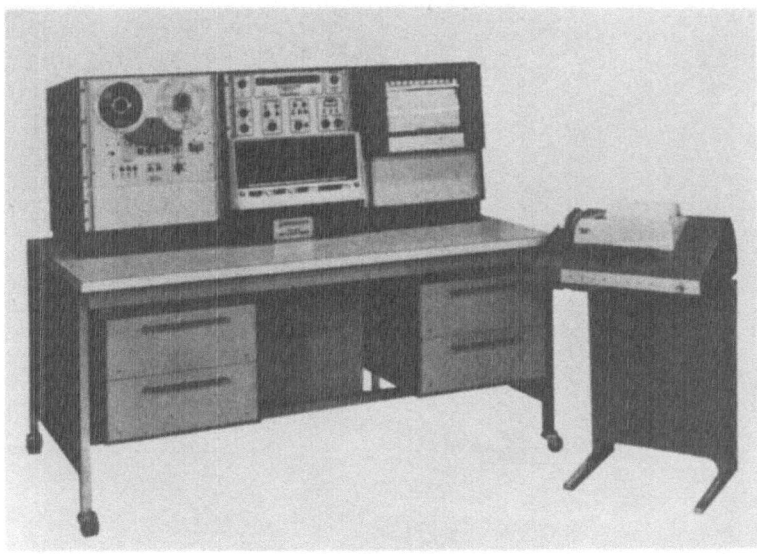

Figure 4. The CRS-40/50 system.

Signals obtained through the voltage-to-frequency converter are recorded as a series of pulses. After this conversion, the only danger to signal purity comes from minor but significant variations in recording speed. To compensate, one can record a constant frequency tone along with the signal and then, as the tape is played back, pick off the tone and monitor its frequency. Deviations from the constant tone can be made to generate compensating voltages to keep the signal stable as it is played into the integrator.

The playback system has one additional advantage: playback of signals can be as much as 16 times faster than recording, thus gaining back some of the time lost by going off-line.

In the case of the CRS-40/50, however, it is possible to have one's cake and eat it too. The system will accept input from an integrator directly, bypassing the tape facilities, and thus can operate either on- or off-line at a rather nominal changeover cost.

If the system *is* to be operated on-line, it is considered good practice to install the integrator at the site of the GC. In this case, a remote control unit such as the one shown in Fig. 5 is provided. This unit, called the RC-110, contains part of the circuity necessary to interface the integrator with the computer. Thumbwheel switches enable the user to dial in appropriate identification data, and all computer-related controls (e.g., computer ready, reset, etc.) are also provided.

Figure 5. An infotronics remote control unit.

Regardless of whether the integrator-computer system is to be operated on-line (via lines) or off-line (in conjunction with tape), a number of devices are available to facilitate analyses.

Perhaps the most fundamental of these is the simple electrometer amplifier. It might be useful, before discussing this amplifier, to briefly review the concept of wide dynamic range (see Fig. 6).

Formally, dynamic range is defined as the ratio of the maximum linear signal to noise level. Graphically, that simply means the peak height shown —about the most that can be hoped for without significant distortion— divided by the tiny noise fluctuations at the baseline.

Chromatograph output peaks vary tremendously in size. Everyone is familiar with the attenuation procedure used with a strip chart to catch all of these peaks. For automatic processing, attenuation switching is out of the question. Nevertheless, the integrator must see the entire analyzer output. Since these systems can accept the total output range of the analyzer detector, which is on the order of $10^6$, a suitable wide-range, high-sensitivity amplifier is required between analyzer and system. Such an amplifier, in the case of ionization detectors, must be an electrometer amplifier and must offer a range equivalent to that of the detector and enable the system to see everything the detector sees without switching. An electrometer designed to accomplish this, the Infotronics EA-1, is shown in Fig. 7.

## 4. PROGRAMING THE INTEGRATOR-COMPUTER SYSTEM

A simplified program flow chart, applicable to composite integrator-computer systems, is shown in Fig. 8. Note that the two central functions— data gathering and data processing, are carried out separately. This is

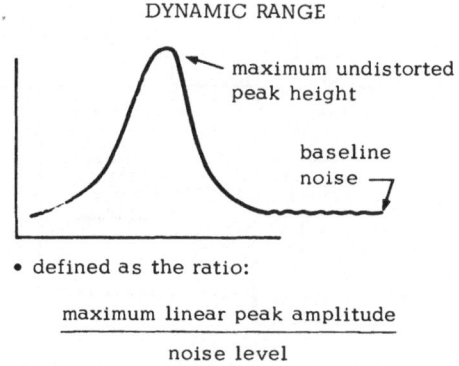

Figure 6. Graph showing the concept of dynamic range.

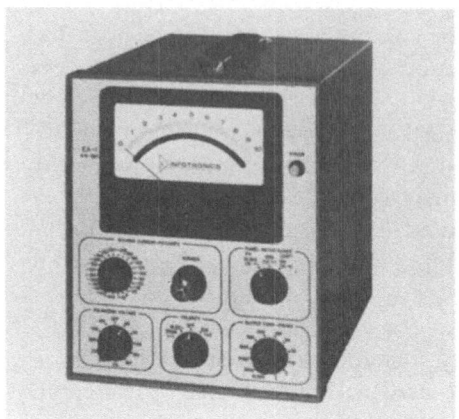

Figure 7. An Infotronics EA-1 electrometer.

handled in the computer on a priority interrupt basis; when the data gathering function is complete for any given run, processing is carried out in the "time windows" between data gathering operations for other GC's.

The program accepts the following types of information: (a) 10 identification numbers; (b) 3 time digits; (c) 7 or 8 data digits in ASCII or decimal form; (d) response factors to four significant figures; (e) run dilution or sample size factor to four significant figures; and (f) names of components.

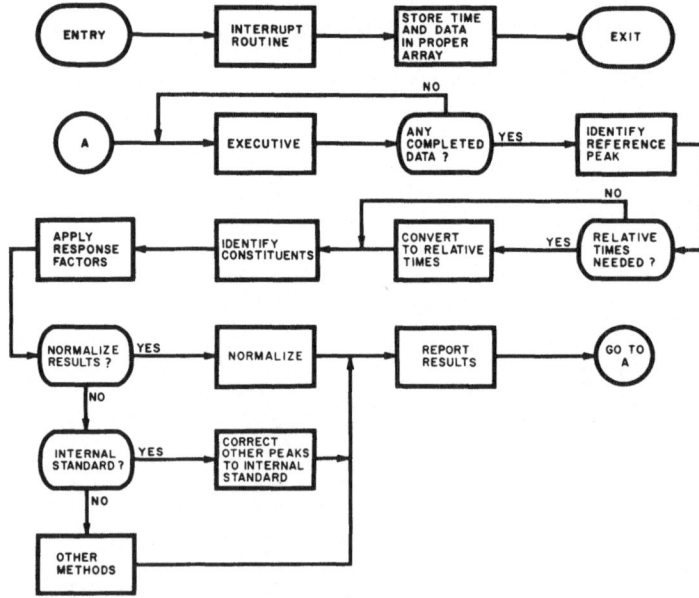

Figure 8. A simplified program flow chart.

Requirements for core space to handle this information, plus incoming data from chromatographs, will vary from system to system. As a rule of thumb, however, the equation below may be used to calculate core space.

$$C = 4096 - 40I + 5N + (4I + 3)P + K$$

where $I$ is the number of inputs, $N$ is the number of names, $P$ is the number of peaks, and $K$ is the amount of program and common memory.

By contrast with computerized data gathering systems which do not employ an integrator's preprocessing capabilities, programing for an integrator-computer system is relatively simple.

The identification, analysis and reporting routines virtually remain a constant size for any size system. The size of the interrupt routines and working storage requirements will naturally vary with the number of integrators in the system. Figure 9 relates the maximum configurations for 4K core memories.

The 4K core system can contain all the program, working storage and tables to manage up to three chromatographs with a maximum of 27 peaks per run.

With more inputs or with chromatographs containing a large number of peaks, the DISK SYSTEM is required. This system, the CRS-50E, is shown in Fig. 10. Data storage routines are changed so that the peak times and areas are stored on the disk rather than in core. In addition, the tables of retention times, response factors and names are stored on the disk. Not only does this configuration handle runs in excess of 100 peaks but it can also be used with up to 10 integrators. As in the core-only systems, the identification, analysis and reporting routines remain the same size in the disk system. Thus, expansion of the system program becomes a relatively easy task.

| | I<br>MAXIMUM<br>INPUTS<br>(MEM.= 40I) | N<br>MAXIMUM<br>CONSTITUENT<br>NAMES<br>(MEM.= 5N) | E<br>MAXIMUM<br>DATA TABLE<br>ENTRIES<br>(MEM.= 4E) | P<br>MAXIMUM<br>PEAKS/RUN<br>(MEM.=(4I + 3) P | K<br>PROGRAM<br>& COMMON<br>MEMORY<br>REQ. | TOTAL<br>MEMORY<br>REQ. |
|---|---|---|---|---|---|---|
| MAGNETIC<br>TAPE INPUT<br>4K MEMORY | 1 | 75 | 150 | 30 | 2800 | 4025 |
| 3 INPUTS<br>ON-LINE<br>4K MEMORY | 3 | 50 | 80 | 27 | 3000 | 4068 |
| 10 INPUTS<br>ON-LINE<br>4K MEMORY<br>32K DISC | 10 | 500<br>(Disc Stored) | 4000<br>(Disc Stored) | 100 | 3300 | 4000 core<br>25000 disk |

Figure 9. The maximum configurations for 4K core memories are related.

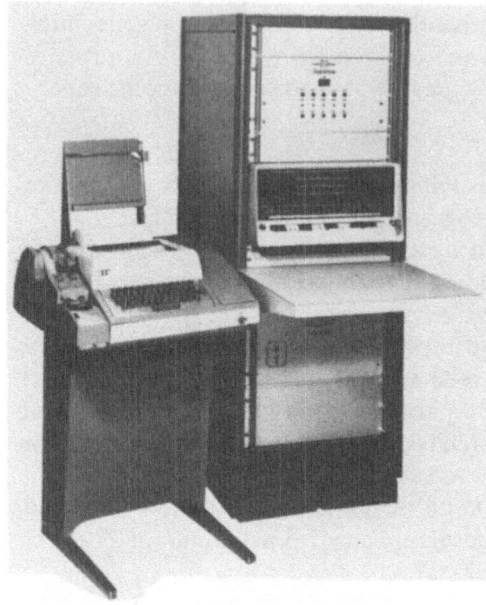

Figure 10. The Infotronics CRS-50E disk system.

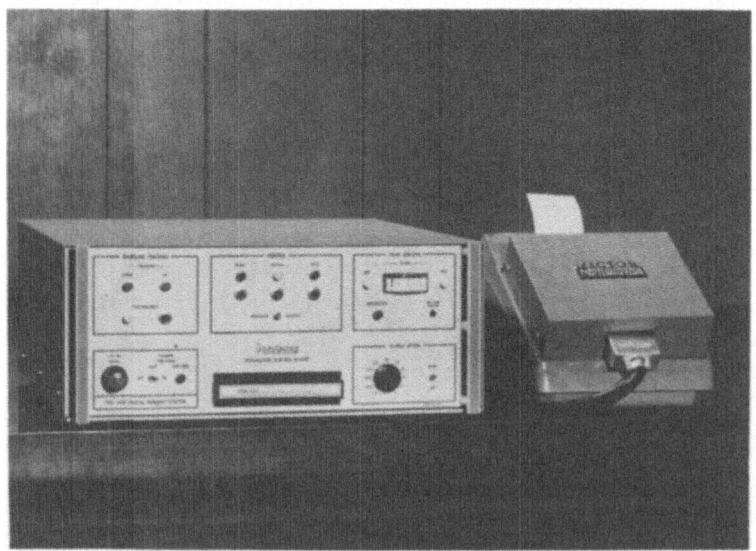

Figure 11. The Infotronics CRS-110A integrator.

## 5. SYSTEM FLEXIBILITY—CONCLUSIONS

Flexibility would seem the key feature of computer-integrator systems. They can easily be expanded or modified to precisely meet the requirements of individual analytical laboratories. It should be noted, too, that they are not necessarily limited to gas chromatography applications. An integrator designed to interface a computer with an amino acid analyzer (the CRS-110A) is shown in Fig. 11. This unit can be teamed with a computer system without modification.

Operating flexibility and minimum down time are equally significant factors in composite systems. If the computer should go down, the integrator can still collect the data that can be entered later. If one integrator is down, another unit can be used in its place. The system can be started with a small initial investment and expanded as desired at a later date.

# VII. ON-LINE OPERATION OF A PE 621 INFRARED SPECTROPHOTOMETER— IBM/1800 COMPUTER SYSTEM

T. Chuang, G. Misko, I. G. Dalla Lana, and
D. G. Fisher

*Department of Chemical and Petroleum Engineering*
*University of Alberta*
*Edmonton, Alberta*

A PE 621 infrared spectrophotometer purchased with shaft encoders and analog-to-digital signal converters was interfaced with an IBM 1800 data acquisition and control system. The real-time computer system programing is currently capable of monitoring and/or controlling a number of such subsystems. The online operation of the spectrophotometer is presently limited to data acquisition and processing.

The infrared encoded wave number and percent transmission signals are resolved to 0.1 cm$^{-1}$ and 0.1 %, respectively. Data smoothing by least mean squares and peak picking programs described by Saviszky were adapted with modifications. Typical spectra will be used to illustrate the performance obtained.

The current application requires differential measurements of IR spectra from adsorbed organic species on various solid catalyst surfaces. The background spectra of gaseous and/or solid phases can be eliminated during the time of measurement; thus providing greater research flexibility.

## 1. INTRODUCTION

This paper describes some of the experience acquired in the Department of Chemical and Petroleum Engineering at the University of Alberta with the start-up and initial operation of a Perkin-Elmer Model 621 infrared spectrophotometer coupled to a real-time digital computer. The experience is that of a group of individuals—some primarily concerned with the extension of infrared spectroscopy into other fields of application and others concerned with developing an IBM/1800 digital computer system with time-sharing capabilities.

One may wonder, validly, how such a facility for the on-line operation of an infrared spectrophotometer was justified. Since the IBM/1800 digital computer was the mainstay of an already existing Data Acquisition, Control, and Simulation Center, it was not necessary to justify the acquisition of such computing facilities with this single application. From the time of its inception, the IR on-line application study seemed to coincide with the goals of the center to provide time-sharing computer operation to students and staff, when needed in their educational, research, and other service functions. With a purchase order for a PE 621 instrument already justified for research needs, it was in essence, an incremental commitment to develop the complete on-line system.

Some of the advantages of real-time computer monitoring of the IR spectrophotometer are:

1. The processing of experimental data by digital methods may be performed during data acquisition, or immediately afterwards, thus eliminating unnecessary intermediate outputs (such as uncorrected raw data). It also allows the IR operator to know that the run was successful and that the results may be used in determining the next step in his research program.

2. Analytical instruments, such as IR spectrophotometers equipped with digital encoders, can generate a great deal more data, and of higher precision, than is normally available with conventional instruments. This additional data and precision is easily handled by the digital computer and, in addition, sophisticated data processing methods, including signal filtering and averaging, can be used in generating quantitative descriptions of spectral bands which characterize molecular species. A number of references in which such concepts are developed are already available ([1,2]).

3. The inflexibilities of a hardware-programed control and data acquisition system are removed. The experimentalist is not limited by an apparatus constructed to perform restricted preconceived functions—the door is always open for simple implementation of new ideas and methods. Descriptions of multiple-user experimental control applications on computers which stress this advantage are now appearing in the literature ([3–5]).

4. Greater research productivity, and even creativity, may accrue from elimination of more tedious data acquisition and analysis functions, from the speedup in flow of processed information, and generally speaking from the greater freedom which may be exercised. The researcher devotes more of his time to interpreting rather than generating results.

5. Once the computer system is programed, it ensures reliable, reproducible analysis and data processing. Problems associated with turnover of technicians (or in the university, that of training new students) are greatly reduced.

6. All procedures used are defined and documented by the programs and all output, files, statistics, etc., are in neat, standard formats.

Once the computerized system is operational it is often almost as easy to do a "complete" experimental run with full computer analysis of the data as it is for the conventional researcher to do a "quick" experiment.

The capabilities of the digital computer in off-line primarily computational applications need not be dwelt upon however, the variety of applications that can be handled "simultaneously". by a real-time time-sharing digital computer is not so well appreciated. To illustrate the versatility of these computers, Table I presents a classification of applications currently being run using the University of Alberta IBM/1800 facility. The IR data acquisition operation can be seen to possess a high-priority interrupt level since failure by the computer to accept data results in lost data.

TABLE I. Classification of Applications at the University of Alberta

| Application | Example | Priority | TSX* |
|---|---|---|---|
| Data acquisition | Experimental data read directly from process inst. at specified intervals. | High (interrupt level) | Skeleton resident |
| Process control | Analog supervisory control or direct digital process control by a general monitor program (real or analog simulated process). | High (interrupt level) | Skeleton resident |
| Specialized research problems | High-speed data acquisition (over 100 pps) with or without data checking. | High to medium | Skeleton resident or V-core resident during execution |
| Data analysis and reduction | Digital filtering and or adjustment for instrument calibrations. | Low if data is from files | Queued mainline program with priority |
| Service programs | Library programs for data manipulation or output: e.g., X–Y plots. | Low | Queued mainline program with priority |
| Background processing | FORTRAN compile; execution utility Programs; Simulation Programs (CSMP). | Lowest | Operate under a nonprocess monitor when computer is not being used for real-time programs |

*TSX: IBM's Executive program.

A block diagram of the IR spectrophotometer interfaced to the computer is shown in Fig. 1. The multi-application capability of the system is also indicated but emphasis is placed upon the participating role of the IR unit. The diagram shows the flow of data originating as the signals,

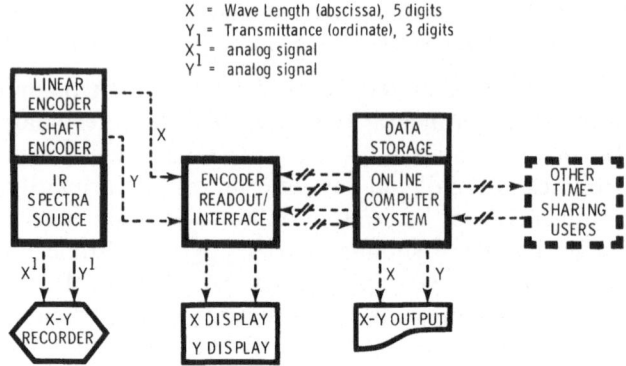

Figure 1. Block diagram of IR computer system.

$Y'$ and $X'$, through to their presentation in several forms of digitized output, $Y$ and $X$. The considerable duplication of function when a conventional hardware-programed apparatus is coupled to the computer is evident: for example, the digitized $Y$–$X$ points, the percent transmission and the wave number, can be read from the IR recorder, the visual display tubes on the interface, or from any of the computer peripherals. In a sense, the first two are superfluous. Also, provision on the spectrophotometer for specifying a scan over a fixed range could be replaced very easily by programed instructions read into the computer.

The advantages of this type of computerized application arise from eliminating conventional input–output control systems, greater quantity and often superior quality of data taken in the same scan time, and benefits accruing from rapid, detailed processing of the data. The latter two capabilities have been available to some extent for some years via off-line computer assistance when optional data-logging equipment, sold by the manufacturers of IR spectrophotometers, is used. In the past, the data acquisition and data processing functions have been physically and usually chronologically separated. Only more recent real-time equipment developments, such as the system described herein, eliminate the separation disadvantages.

On the other hand, the "stand-alone" capability of the IR unit is still desired in many problems, especially ones which are unanticipated and which arise frequently in research activities.

## 2. SYSTEM COMPONENTS AND THEIR OPERATION

This section will describe in more detail the three major equipment items comprising the PE 621-IBM/1800 system, and their method of communicating information or instructions.

## 2.1. PE 621 Infrared Spectrophotometer

This instrument and its capabilities are well known and so, detailed comments on its design are unnecessary in this report. The conventional output signals are percent transmittance ($Y'$ = source signal, $Y$ = digitized signal) and wave-length ($X'$ source signal, $X$ = digitized signal). The PE 621 was purchased with Perkin-Elmer $X'$ and $Y'$ encoders installed. The two encoders are one-brush absolute position types and their shaft positions are tracked continuously by flip-flop registers ([6]). Freedom from gearing, backlash and possibly servo errors, particularly those inherent in systems taking data from the recorder servo unit, may also be eliminated. The $Y$ signal in three digits is obtained from the comb servosystem. The $X$ signal resolved to a tenth of a wave-number, is obtained from a 100-count encoder mounted to the wavelength shaft (monochromator). An up–down counter (three digits) in the interface unit counts the number of revolutions of the 100-count encoder (two digits), thereby giving a total wavenumber of five digits. Since subsequent transmission of the encoded signals to the computer is in BCD code, this system permits the monitoring of $Y$–$X$ signals with high resolution, the resolution being limited only by that of the encoders.

The operation of the PE 621 is presently initiated by operator control however, future plans include a modification to permit initiation by computer command.

## 2.2. Encoder Readout-Computer Interface

The Perkin-Elmer Model 8RLS encoder readout-computer interface includes an operating mode for direct transmission of $Y$–$X$ data to a digital computer. The signals are received continuously from the IR in encoder switching code and are then translated to BCD and stored in an encoder register. The BCD data may be decimalized for visual Nixie display or alternatively, transmitted to a data-logging device such as an external paper tape punch coupler. The data in the encoder register is strobed into a buffer storage register which freezes the information. A translator converts the buffer register output into 8-4-2-1 BCD output in positive logic (logical one = $+8V$, zero = $-12V$) and transmits it to a rear panel connector for direct hookup to the IBM/1800 computer.

The encoder readout has a maximum capacity of 32 bits (eight decimal digits) and can accept up to 300 bits/sec (roughly, ten $Y$–$X$ points/sec). The data encoding rate is operated asynchronously with the computer since synchronized real-time monitoring of data points is not desirable with the hardware-controlled operation of the spectrophotometer. The wave number interval which determines the data sampling rate is specified by operator selection of multiples of the minimum increment ($0.1 \text{ cm}^{-1}$) available for the $X$-axis. The sampling rate is generated by a data trigger which strobes the signals at the specified intervals of $X$ from the encoder register into the buffer register for parallel transmission of all bits in the 8-digit data word to the computer. Each time that a $Y$–$X$ point has been transmitted from the encoder register to the buffer register, and thus to the computer, the encoder register is left in a "reset" position. The computer acknowledges receipt of the data point by sending a voltage pulse to the interface unit (one wire pair needed), which returns the encoder register to "set" position. If a second point should be strobed into the buffer register without the preceding computer acknowledgment, the "reset" position indicates loss of data and initiates an alarm condition.

The computer and IR-interface are separated by roughly 125 ft of communication cable (two 18 pair 22-gauge individually shielded twisted-pair cables). The cables are totally shielded as well and are run through conduits for mechanical protection, providing rather excessive shielding protection. The BCD signal, the start-up communication, and the computer acknowledgment pulse require 34 pairs leaving two pairs in reserve. It is interesting to note that the essentially noise-free 32 pairs could be replaced by at most two pairs if the "analog" signals, $Y$–$X$, were to be transmitted. Noise pick-up in transmitted $Y$–$X$ signals is more likely and could reduce the data accuracy considerably.

The direct generation of digital values and the transmission of the digital BCD signals should be essentially noise-free. The quality of data input to the computer therefore is directly dependent on the accuracy of the encoder readout and the basic IR components.

## 2.3. IBM/1800 System

The hardware configuration most pertinent to the on-line operation of the IR unit is shown in Fig. 2. The equipment designated by the boxes with slightly thicker margins has been involved in the current studies. Access to such an extensive facility with its large core and disk storage facilities, full range of conventional input–output peripheral hardware, substantial computing power, and supporting systems analysts greatly simplified the task of developing a powerful and convenient-to-use system.

To illustrate the systems role, the operational capability of the IR project was designed to proceed via three general phases:

1(a) On-line data acquisition and conversion, under program control, using a specialized IR program which placed the data in FORTRAN accessible disk files.

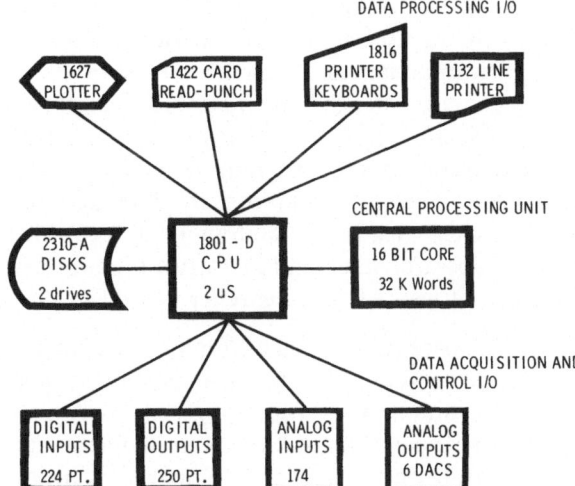

Figure 2. IBM/1800 data acquisition and control system.

(b) Development of suitable programs to analyze the data and output the desired graphs and printed data on the central printer and/or plotter or on remote terminals and/or storage oscilloscopes.

2(a) Similar to 1(a) except that data channel operation with external synchronization could be used thereby reducing the computing and interrupt servicing load. Programs could be written to "overlay" part of the "permanent user's area in core" when the IR run was initiated.

(b) Interfacing of the IR spectrophotometer to permit real-time control of variables such as wave number, sampling rate, attenuation, etc.

3. Similar to 1 and 2 except that the IR unit would be one of several input devices served by generalized data acquisition and control monitor programs which would always be available for immediate execution.

4. Use of real-time communication facilities to the University's IBM 360/67 system to facilitate extensive data processing and/or file searching such as would be required for compound identification.

The system is presently operating with all of the phase 1(a) programing permanently resident in core. As a result, the initial programing and debugging is greatly simplified but this retention of core is inefficient when the IR system is inoperative and other users' core requirements cannot be met. Although on-line operation has been successfully accomplished and the data processing seems to be free of major programing errors, additional work on phases 1(a) and (b) is still required. Phase 2(b) and 4 will be implemented by the user as further experience is gained, or when their physical feasibility materializes. Implementation of phases 2(a) and 3(a) will be left to the computer systems analysts since they are concerned with computer utilization and these changes do not affect the user directly.

Extension of the program to include computer control of the IR spectrophotometer in addition to data acquisition as outlined in 2(b) would "close the loop" and make it possible to perform experiments without operator intervention. Advantages of this type of operation from the point of view of the researcher as well as the computer include:

1. IR parameters such as attenuation can be changed more flexibly by the computer during a scan as a function of the current or real-time conditions thus improving sensitivity and/or accuracy.
2. The operation of the IR instrument and associated experiments can be monitored and compared against a mathematical model or specified tolerances so that malfunctions or failures (such as omission of an expected absorption band, a change in the transmittance thermocouple detector, burnout of the IR source, etc.) could be detected and brought to the attention of the operator, and/or corrective action initiated by the computer. Detection and notification of unexpected phenomena, at the time they occur, usually permits the operator to determine the cause more easily and may point out areas for improved methods or new research.
3. Filtering and/or averaging can be combined with on-line data acquisition to offset the effects of noise. Computer control would make it possible to leave the decision regarding the need, and the implementation of rescans of portions of the spectra to the computer. Similarly, data-sampling rates could be varied as a function of, say, the value and rate of change of the transmittance and the resolution desired.
4. Computer control of the incrementing of the wave number could mean that the IR instrument responded to control commands from the computer instead of requiring the computer to respond to high priority interrupts from the IR. This would make time sharing of the computer more efficient and reduce interrupt conflicts.

The IR data acquisition program referred to in 1(a) responds to interrupts generated by the interface unit and reads in the 32 bits of data. The 32 BCD input data bits are converted to binary integer numbers and stored into two 16-bit words of a core buffer. When the 320-word core buffer is filled, it is transferred to one sector of disk storage. In view that this is a time-shared operation, the disk could be busy completing a disk operation for another program. Thus provision had to be made for queuing time and disk-seek operations as well as for the actual 10 ms. of write time. A 40-word auxiliary core buffer is provided to collect data during the disk operations. At the maximum data rate of approximately ten samples per second this provides 2 sec in which to complete the entire disk operation. The size of the auxiliary buffer can be changed by an equate card at system generation time.

The disk data files were stored in a form accessible by FORTRAN coded programs so that the users could write their own data processing programs. Since the five-digit wave number could be as large as 40,000 and since the largest number that can be represented by the 16-bit data word of the IBM/1800 is 32,767 it was necessary to use either a two word integer or a special data code. To conserve core, it was decided to store the four least significant digits of the wave number in a one integer format with the associated transmittance in the adjacent word. Each time that the fifth or most significant digit changes, it is recorded by putting a special entry in the file with a transmittance of "−1" (a physical impossibility) and a wave number equal to the fifth, highest order, digit. The users' program can then access the file and handle it as one-word integers, or convert it to two-word integers or floating-point numbers. Negative values for percent transmittance are also used to "flag" special codes to indicate run termination or alarm conditions.

## 2.4. Operation of the System

The flow diagrams in Figs. 3–5 describe in simplified terms the role of the computer during the startup, operation and shutdown of program-controlled data acquisition. The IR operator in the remote laboratory actuates a "request to computer" electronic signal when an IR scan is to be performed.

In the *start* procedure, upon assuming the "ready" state, the computer informs its operator via print message that it has initialized data acquisition and has also informed the IR operator of its "ready" state. The latter is

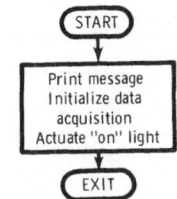

Figure 3. Encoder-computer flow diagram
Phase 1A: program controlled data acquisition
(a) start procedure.

IR operator initiates start via push button
Exit implies return of control to TSX system
for processing of other programs.

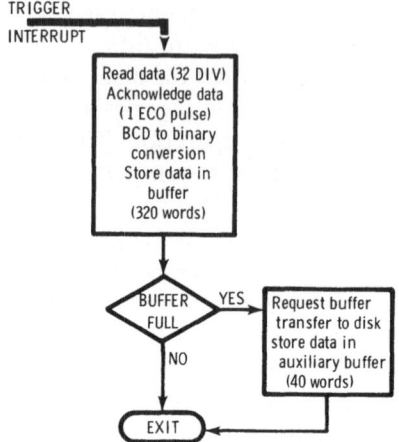

Figure 4. Phase 1A: (b) data acquisition.

accomplished via a light on the panel of the interface unit. It then returns to its time-sharing service role until the IR operator, at his convenience starts the hardware-controlled IR scan.

When reading data, the computer files all of the raw data before any data processing is commenced. When the IR operator signals by turning the on-line switch to the off position that the scan has been completed, the computer again advises its operator that the job has been completed and it then summarizes the system statistics.

## 3. DATA PROCESSING

The treatment of infrared spectroscopic data which has been obtained with the system described herein involves processing up to 35,000 $Y$–$X$ points if a full range of 3500 cm$^{-1}$ is scanned. These points would occupy 70,000 words of memory, or more, if several complete spectral scans were to be retained. Only with access to a time-sharing computer system, with relatively large disk storage and rapid computing capability, can the

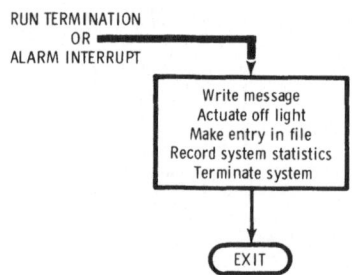

Figure 5. Phase 1A: (c) stop procedure.

experimenter obtain reduction of such large amounts of data in a speedy manner.

Data processing routines such as data filtering, peak picking and other numerical methods can be used on the data file, and where advantageous, the benefits from varying the computational parameters can be exploited. The raw data and/or the resulting smoothed data can be stored on disk cartridges or punched cards, printed on the line printer or typewriters, or displayed on the digital plotter or the oscilloscope. Present plans for this DACS Center call for an on-line link to a large central computer so that, for example, if the need arises the listing of mid-peak frequencies could be transmitted to the large computer for addition to, or comparison with, library files for known molecules.

For the sake of efficiency, that of both user and computer, the development of adequate data reduction programing must involve a careful evaluation of the user's needs. Smoothing of the data is advantageous in research applications such as those to be described if spectral characteristics may be ambiguously interpreted because of high random-noise levels in the percent transmittance or $Y$ data, i.e., random error superimposed upon the significant output signal. If the signal-to-noise ratio is very high, routine smoothing of data may be wasteful through unnecessary use of computational time in a time-sharing system. If selection of peaks, i.e., the significant spectral bands which characterize the sample molecule, is required, then some level of smoothing of the data may be advantageous so that those changes in slope which define significant peaks can be more readily detected. A considerable amount of work on such problems has already been reported by Savitzky. Much of this thinking and programing ([7]) has been incorporated into the data reduction methods employed in this work. The methods involve least squares fitting and assume that the error is associated with the $Y$ signal (three significant digits) and not with the $X$ signal (five significant digits).

Figure 6 presents a simplified data-reduction flow diagram which relates the major steps involved in the data smoothing and peak-picking routines. The data acquisition involves multiple-user on-line operation and so, the data reduction and output functions are not commenced until the scan is completed. Although not shown in Fig. 6, the data points are read from the disk file, sector by sector. They are subjected to least-squares smoothing and a slope calculation except where the transmittance reads $3\%$ or less. The slope is calculated at each point using a quadratic function fitted to the smoothed data. When the sign of the slope inverts, defining passage across an absorption band, a 15-point cubic equation is used to determine the ordinate and abscissa at the peak minimum. The raw data is completely treated, the smoothed data replaces it in the disk file. At present, if a plot of the raw data is required, this output must precede the smoothing

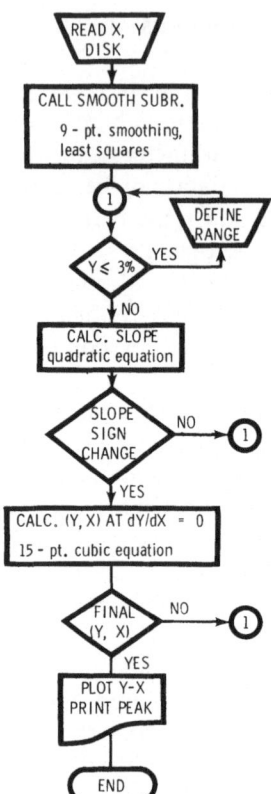

Figure 6. Data reduction flow diagram.

since the latter is destructive to the original memory file. Upon completion of these steps, the *Y–X* plotter programs calculate appropriate scales and automatically plot the function. The evaluation of the effect of computational parameters upon smoothing and peak-picking, e.g., different levels of filtering, was usually done by off-line processing.

Figure 7 illustrates a comparison on an expanded scale between some raw and smoothed data for a polystyrene film in an ATR cell. The figure was actually plotted by programed control, in the manner shown, to conform to a preset 18-in. ordinate scale. This range of the *X*-axis was chosen because of the relatively high noise level visible on the PE 621 drum recorder chart, a line of thickness roughly 1 % transmittance. The four graphs on Fig. 7 show the raw data, taken at the highest sampling rate of every $0.1$ cm$^{-1}$ interval, and the influence of smoothing this raw data. Some differences in data filtering are indicated from using all of the data, half or one-quarter of the data.

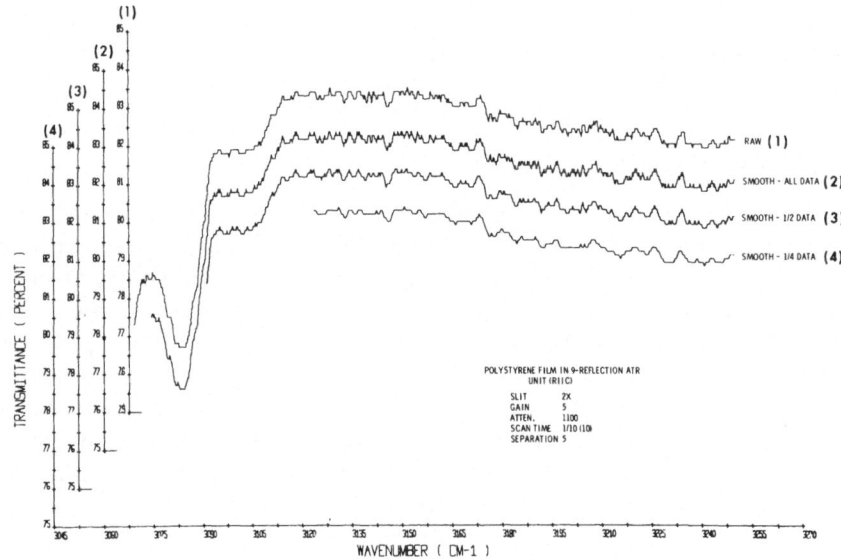

Figure 7. Comparison of raw and smoothed data.

In the smoothing procedure, it would be expected that the nine-point filter band (or "smoothing gate") is effective in suppressing random noise if the width of the gate is considerably larger than the period of the noise frequency to be removed. The application of smoothing to the complete file of data sample involved a 0.9 cm$^{-1}$ gate-width (as shown by the second plot of Fig. 7); this did not prove to be effective. Since the wider spacing of the data points produced by using every $N$th point in the file generates a wider effective band with the nine-point filter, the smoothing became more noticeable as can be seen on plots 3 and 4. Because less data is employed when the gate is increased, the quality or significance of the resulting plot could also be reduced. Replication of a run, and averaging to eliminate random noise, is essential to establish the best smoothed function which could be used as a criterion of smoothing effectiveness. This standard was not available at this time for comparison with the plots of Fig. 7. This phase will be undertaken shortly and the results, if of general interest, will be reported in the future.

An alternative explanation, originating with Savitzky, relates the lack of smoothing which resulted from use of all of the data to the difference in magnitudes of time constants between that of the signal-generating instrument and that associated with the span of smoothing-gate. Until the time constant of the latter exceeds that of the instrument, it would be unrealistic to expect beneficial smoothing. This may be the case for the ATR signal until the time constant associated with a $1.8\,cm^{-1}$ gate span was attained. This span was obtained with the nine-point gate using each second $Y-X$ point.

## 4. RESEARCH APPLICATIONS

The research programs planned or underway involving infrared spectroscopy and which will utilize on-line capabilities have presented two types of data interpretation problems. The first involves a comparison of infrared spectra (with or without noise) recorded "before and after" the experiment to determine changes which have occurred. The second problem involves repetitive studies in which both an external experimental parameter is varied and then an IR scan is performed until a detectable change in spectral character emerges. Both of these studies should profit from computerized treatment of data: the former, by more sensitive diagnostic analysis resulting from increased numbers of data points and by elimination of tedious point-by-point visual measurements, and the latter, by replacement of manual control with direct digital control permitting on-line sequential experimentation when the control logic can be physically implemented. Both problems are to some extent limited in their scope by the hardware-programed control of the IR unit. On the other hand, because the IR unit is designed for "stand-alone" capability, this feature enables the operator in attendance to change experiments or conditions quickly when nonroutine unanticipated research results are encountered.

### 4.1. Measurement of Differential Spectra for Experiments on a Single Sample

Figure 8 illustrates some of the spectral characteristics which are encountered during infrared studies of adsorbed molecules and surface reactions on solid catalytic substances using the Eischens technique ([8]). By plotting the difference in percent transmission between the two spectra at the same wave number, taken before and after the experiment and noting also the mid-points for all peaks, the extensive data is reduced to that which appears to be significant.

Often, the full spectral range spanning $3500\,cm^{-1}$ must be studied for the before as well as the after experiments, so that the significant differences between the two sets of spectra can be plotted. The differences between the two sets of spectra may be plotted in the form of the "delta-plot" shown in Fig. 8. This being a tedious but simple task it is ideally suited for computer processing, especially when smoothing is advantageous. The delta-plot shows only the differences in transmittance at each wavelength and thus classifies the changes according to whether they are positive or negative

Comparison of IR spectra "before and after" the experiment.

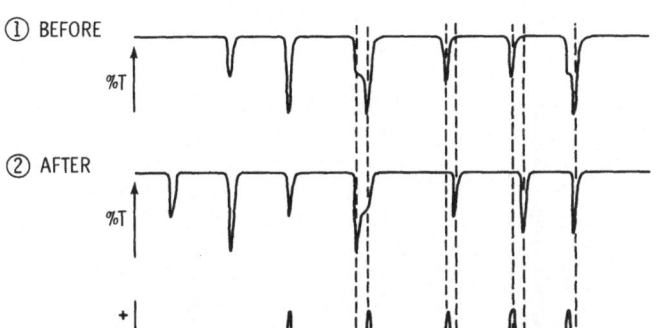

Characterization of band differences by " Δ-plot" plus printout of mid-peak wave nos.

Figure 8. Research applications I.

peaks. Positive peaks imply an increase in an existing band or the appearance of a new band, except when a frequency shift occurs. The negative peaks indicate either disappearance or a decrease in transmittance for an existing band. The use of the delta-plot together with a list of mid-peak frequencies for both before and after spectral scans, all provided by computer output, would represent savings in calculation time and an improved reliability in the results.

Since smoothing is influenced by computational parameters such as sampling rate and width of the smoothing-gate, a replication test should be carried out early in each research program at conditions representative of those to be encountered. The replication, i.e., averaging of a number of repetitive scans to eliminate random error, provides a standard, the duplication of which should be the objective of any good smoothing procedure. Reference to this standard should enable one to determine whether or not smoothing is advantageous. If it is, one can evaluate the effect upon smoothing of the various parameters, enabling all subsequent similar experimental studies to employ optimal filtering of noise.

## 4.2. Infrared Monitoring of a Reaction System

Figure 9 illustrates a proposed scheme whereby the flexibility and reliability of a high-quality infrared spectrophotometer can be used as an "infrared analyzer." The experimental parameters, such as temperature or reactant flow-rate, are under direct digital control during either a planned

On-line monitoring and control of chemical reaction/kinetics studies.

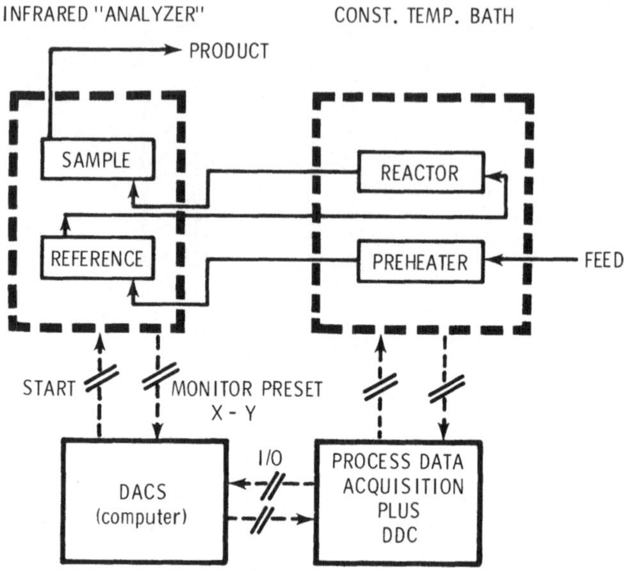

Figure 9. Research applications II.

sequence or computer-determined sequence of experimental conditions. The IR spectrophotometer monitors a particular band continuously via the encoder readout-computer interface or alternatively, it is instructed by the computer to commence a scan over a preset range. The latter scan may be repeated as often as demanded by the computer. Both qualitative and quantitative analyses are feasible with such a scheme.

## 5. COMMENTS

The use of computer processing of infrared spectral data is not a closed subject nor is it fully documented, especially concerning the additional problems or benefits inherent in the transition from off-line computer data processing to on-line data acquisition and control. Generally speaking, much of the activity dealing with IR data processing, either off-line or on-line, has been justified through the increased availability of spectra of improved quality. The hardware and programing to acquire such spectra are available today. With an increased need for on-line capability, a gradual increase in conversion of the important hardware-controlled IR spectro-photometer functions to software-controlled ones may be anticipated.

If on-line capability appears to be advantageous in an application, the time-sharing computer approach seems to be the logical vehicle for its implementation. The principal arguments for a time-sharing system include: (a) access to a larger computer system than could be justified for the single application; (b) the flexibility by which applications can be changed or expanded; and (c) the efficiency that arises from being able to use the computer system for other applications when a particular experiment is shutdown or is inoperative.

The unique set of circumstances at Alberta, which required only an incremental investment of approximately $7000, for an interface and hook-up, have enabled the on-line infrared spectrophotometer operation to be commenced. Although the system is operational, it is currently in an early phase of the multi-application time-sharing program. The bulk of the experience reported in this article deals with the method of operation and the interaction of the IR unit with the system. The objectives, which hopefully will be attained as a result of the two research studies described, are to show that qualitative and quantitative identification of phenomena or chemical species can be improved by computer processing of IR data and to show that "immediate" accessibility to these improved results provides a distinct advantage in situations involving sequential experiments.

A self-evident truth, occasionally forgotten, encountered in computer applications is the requirement that the scientific objectives and logic be clearly specified and thus capable of being programed. If this requirement is met, the goal of preparing adequate IR data reduction software should only pose relatively minor problems today. On the other hand, it should also be noted that these problems, though minor, can be very time-consuming, e.g., debugging of trivial programing errors.

A comment on the experience at Alberta, which involved an interdisciplinary group comprising a chemist, a chemical engineer, a systems analyst, and a systems engineer, might be worthwhile. The minimum staff-time to be expended in such a project is of the order of:

1. Three to four man-months of a B.Sc.-level chemist's time, assuming he has no prior FORTRAN training and negligible IR experience;
2. One man-month of systems analyst time for preparation of the necessary supporting system software, plus consultation on application programing;
3. Occasional supervisory or policy decisions by the two senior staff members, extending over the above periods.

The actual project duration extended over six to eight months excluding major shutdowns. It was also found to be extremely advantageous to have a second member of the staff directly involved in the applications programing and IR operation as an emergency replacement for the chemist. This proved to be very valuable, when a personnel change occurred, enabling the project to continue without serious interruption.

The system software is very formidable, requiring man-years of programing time for a time-sharing operation. It is safe to say that without access to an existing computing center and its staff, this limited exploratory project in the application of infrared spectroscopy would still be an idea.

## ACKNOWLEDGMENT

The assistance of the DACS Center staff at the University of Alberta, and in particular Mr. T. McManus, in devising systems programing relevant to this on-line study is acknowledged. Equipment grants from the National Research Council of Canada and the University of Alberta, enabling the purchase of the PE 621 IR and Interface Units, are also acknowledged.

## REFERENCES

1. A. Savitzky and J. E. Golay, *Anal. Chem.* **36**, 1627 (1964).
2. A. Savitzky, paper #4 presented at the 8th European Congress on Molecular Spectroscopy, Copenhagen, Denmark, August 17, 1965.
3. D. Secrest, *Ind. Eng. Chem.* **60**, 75 (1968).
4. G. Lauer and R. A. Osteryoung, *Anal. Chem.* **40**, 31A (1968).
5. C. H. Sederholm, P. J. Friedl, and T. R. Lusebrink, paper presented at Pittsburgh Conference on Anal. Chem. and App. Spectroscopy, March 1968.
6. Instruction Manual for Model 8RLS Encoder Readout/Computer Interface, The Perkin-Elmer Corporation, August 1967.
7. Adopted from FORTRAN routines supplied by the Perkin-Elmer Corporation to users, which were written by A. Savitzky (P.E. Corp.) and by R. A. Crisler (Proctor–Gamble).
8. R. P. Eischens and W. A. Pliskin, Actes Congr. Intern. Catalyse, Paris (1960).

# VIII. D . . . COMPUTER, WHERE'S MY CURVE?

## W. R. Kennedy

*American Cast Iron Pipe Company*
*Birmingham, Alabama*

While statistical regression fitting of spectrometer curves may seem to be the best way to locate your working curve using a computer, statisticians agree that the exact equation for the curve is unsurpassed in relating the correlation of two variables. The concept and development of such an equation for Baird spectrometer curves are presented, which involves use of only four selected data points to solve for five steering constants, one of which is always known for a given system. The discovery of simple basic relationships at the reversal end and background end of the curve has eliminated the need for superfluous samples and regression fitting.

Papers this year which describe computer programs to develop coefficients for regression fit of instrument data have been numerous. Such programs, or software (in the computer jargon), can become rather lengthy for polynominals of the fourth order and higher. Except for hardware capacity, and software length, development of the necessary coefficients is as easily done for eighth order as for third order—as far as the spectroscopist is concerned.

As the order of the polynominal increases, however, the minimum number of necessary data points increases. No statistician would recommend using minimum data for this type curve fitting. The errors are just too liable to give questionable coefficients. Even with adequate data, regression analysis involving polynominals can only approximate the position of the curve; a correlation coefficient of one is never achieved. The fallacy, of course, is that if the working curve is not a polynominal, then polynominal fitting will never be anything but a fair guess and limited to the extremes of the data used. In my own case, and this I find most important, when considering a fourth-order polynominal with five coefficients, I am concerned that none of the coefficients has any theoretical meaning as far as the spectrometer curves are concerned.

You really cannot appreciate the value of knowing the relationship between—say, percent and intensity ratio, or, as in the Baird instrument, percent and clock time—until the D . . . computer looms in sight. Computer people snoop around, asking truly embarrassing questions to which there are no intelligent answers. This is not a one way street though! I have found the computer people have some standard unintelligent answers, also.

I discovered about five years ago that computers were here to stay, however, my quandary was: analog or digital? I had built a simple one-channel computer to hook to the D.R. ([1]), which would present a voltage reading equal to element percent, covering three decades of the log percent. (Credit goes to J. A. Norris, as he built one first in 1959 ([2]); my version was all solid state and involved no mechanics.) Unfortunately, this computer liked to follow a straight line curve all the way, and while my curves were dynamically corrected for background (making the lower two decades straight), the high end in some curves showed reversal. The computer, in its prototype stage, had no provision for changing slope dynamically, so it was unusable, except for those curves with no reversal, or portions of curves out of the reversal region.

Abandoning this approach of slaving an analog computer to the Baird readout, I had the idea, that with operational amplifiers, it should be possible to use the phototube currents in a log–log ratio system to indicate element percent directly as soon as the shutter opened. While this version worked with simulated currents, it would not operate in the proximity of the D.R. because of the surrounding R.F. field. I have abandoned this approach from immediate plans, but not from my thoughts. It is an intriguing approach and will work, but will not be discussed further here.

They say that sometimes if you procrastinate long enough, the problem will go away. And so it was in my case; the decision of which approach, analog or digital, was settled when our management decided digital! So the D . . . Computer is coming. I suspected this sometime before the decision was made and managed to acquire a small Ollivetti 101 Computer to perform some calculations we were already doing too slowly.

Although this computer is slow, small, and memory bound, it has a number of advantages, not the least of which is availability. Use of our IBM/360 is almost by appointment and at inconvenient hours. With the Ollivetti, I was able to develop workable programs in a short time which ultimately gave the general equation for the spectrometer curve. The concept of this equation and its gradual development is what I would like to describe to you now. The complete program for development of the steering constants has subsequently been written in FORTRAN.

To begin, the Baird capacitor discharge system output is a time duration pulse. I point out here that where the time, $t$, appears in the equation, you may substitute log $I$ ratio and the equation will still be applicable to spectrographic curves.

If we charge two capacitors with constant current, one always to a higher voltage than the other, and place the two in series and then place a bleeder resistor across the high capacitor voltage and measure the time it takes for the algebraic sum of the two capacitors to be zero, the log ratio of the two voltages will be a linear function of time. In other words, in Fig. 1, log voltages are plotted against linear time and the result is a straight line of $-RC$ slope. The $RC$ is the product of the bleeding resistor (ohms) times the bled capacitor (farads). Using pure currents to charge the capacitors, this is the only curve you can attain; this is our anchor in the equation we are developing. This slope $-RC$ is a fixed quantity for a given Baird instrument. We measured this quantity very carefully in our instrument and found, in the eight possible reference capacitors, we had two slightly different $RC$ slopes. We call this slope the electronic slope and label it $M_B$. Its units are log percent per clock division. Figure 1 indicates how the value of the slope is obtained.

Since none of the currents in the spectrometer is a pure current, the voltages which arise on the integrating capacitors are not truly representative of the signals we hope we are measuring. Returning to our two charged capacitors: if we add a small constant voltage to the smaller capacitor, but discount it in the plot, then Fig. 2 shows what happens. In Fig. 2, 0.5 V has been added to the small capacitor. Note, e.g., when the theoretical indicates 2.5 V, we have plotted 2 V. In like manner, the solid curve

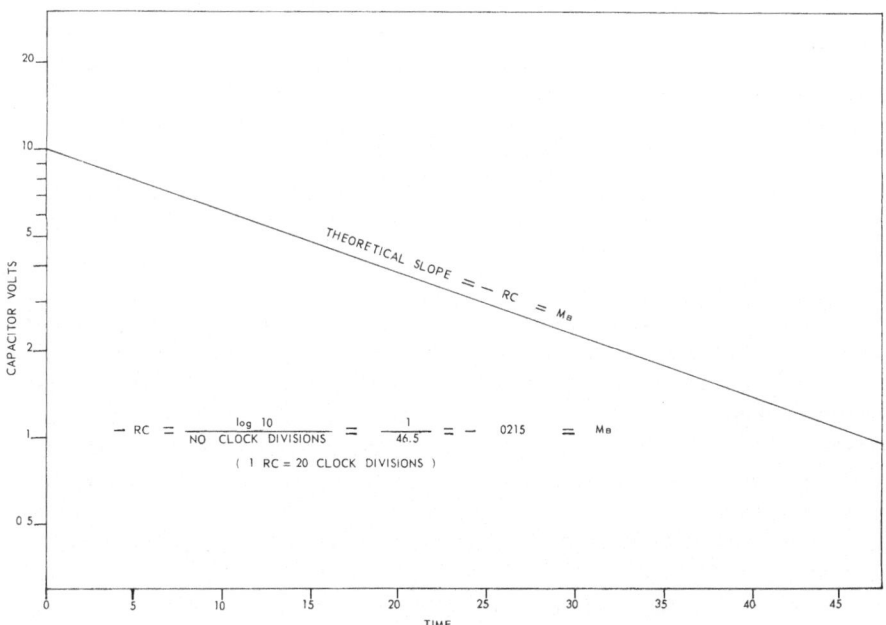

Figure 1. Log voltage *vs.* time.

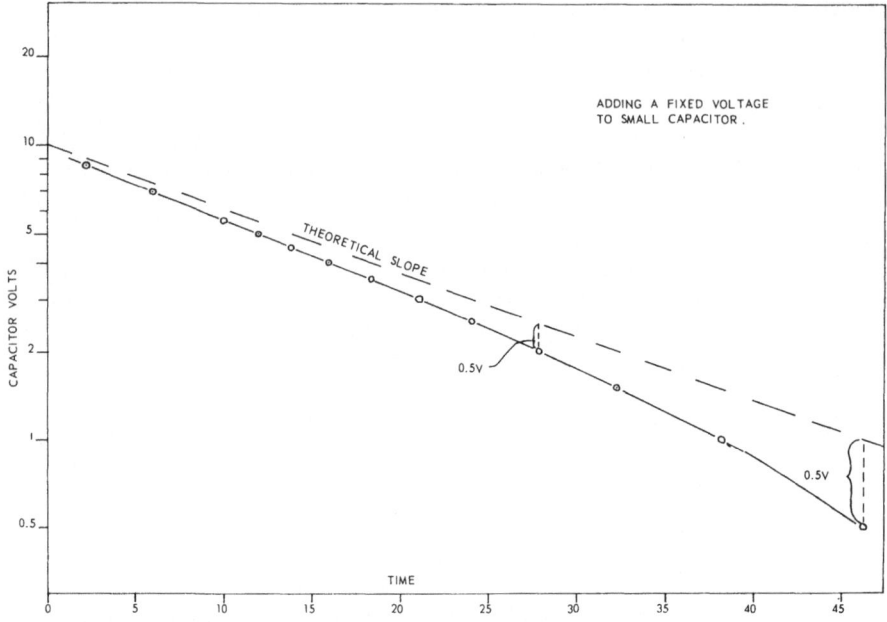

Figure 2. Uncorrected capacitor volts *vs.* time.

represents a constant 0.5 V difference below the dashed, theoretical curve. This is analogous to charging the capacitor with signal-plus-noise and plotting only signal *vs.* time (isn't this what almost everybody does?). In 1963, we built a background correction system to effectively electronically remove the background (or noise) portion of the capacitor voltage from the capacitor containing signal-plus-background voltage ([2,3]). This essentially gave us back the straight line plot on a dynamic basis. We still use this system on a routine basis.

Looking at Fig. 2, I reasoned that if the curve had a downward bend due to the relatively constant background signal on the capacitor, then the curve could be corrected mathematically for background, *if* I knew what percent background to subtract from the theoretical curve. This would be a constant percent all the way up the curve. The immediate solution became apparent, and the Ollivetti computer was used to arrive at the answer as shown in Fig. 3.

Figure 3 shows the lower portion of a curve with two points labeled $B_1$ and $B_2$, which are standard samples running at times $T_1$ and $T_2$, respectively. With the computer the electronic slope was drawn through $B_1$ giving the log intercept of this curve. Since we have to subtract something constant from this straight line to make $B_2$ coincide, it is obvious that a

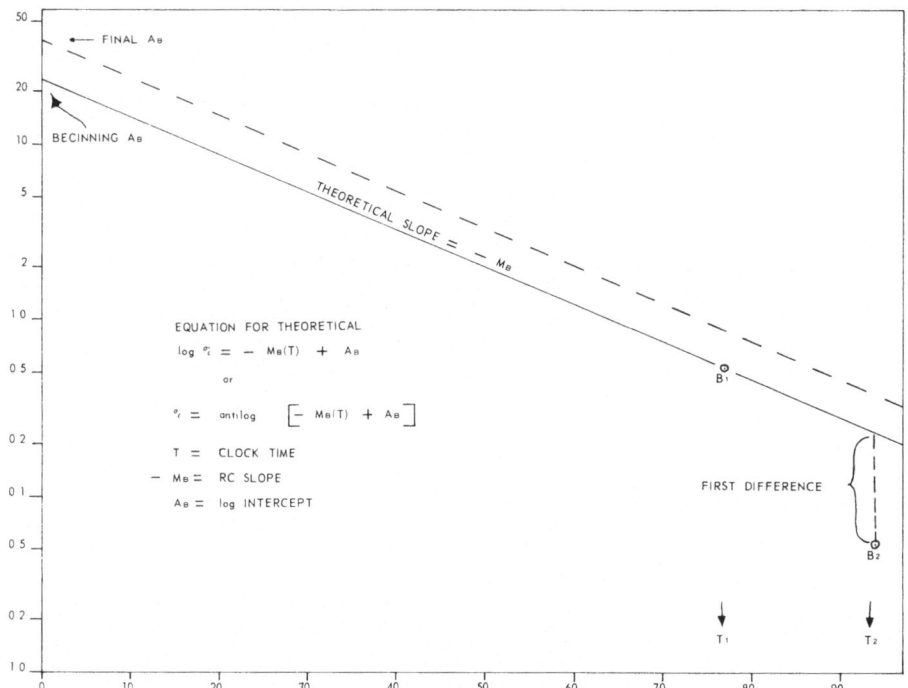

Figure 3. Means of obtaining percent background.

good place to start is with the difference $B_2$, from what $B_2$ calculates to be on the straight line above it. If we subtract this same percent from $B_1$, then this new value would be below the $B_1$ standard value, a negative quantity. So it follows that all that is necessary is to move the straight line upward by increments (incrementing the log intercept) until the two differences at $B_1$ and $B_2$ are equal in percent.

The Ollivetti does a nice job of moving the curve, printing the last log intercept and the percent background to subtract. I carried the mantissa to the nearest ten thousandth because of the computer slowness, but with the IBM/360 in FORTRAN, incrementing the log intercept with five "do loops," to cover the change in mantissa by tenths, then hundredths, etc., to hundred thousandths, cannot run more than 90 times in the last four places. The last log intercept is called $A_B$ and its units are log percent.

Figure 4 shows us where we are by applying increasing clock times to the equation we have so far, of:

$$\text{antilog}[M_B(t) + A_B] - \% \text{ Background} = \% \text{ Element} \qquad (1)$$

This curve is now corrected for background all the way up the curve. If there

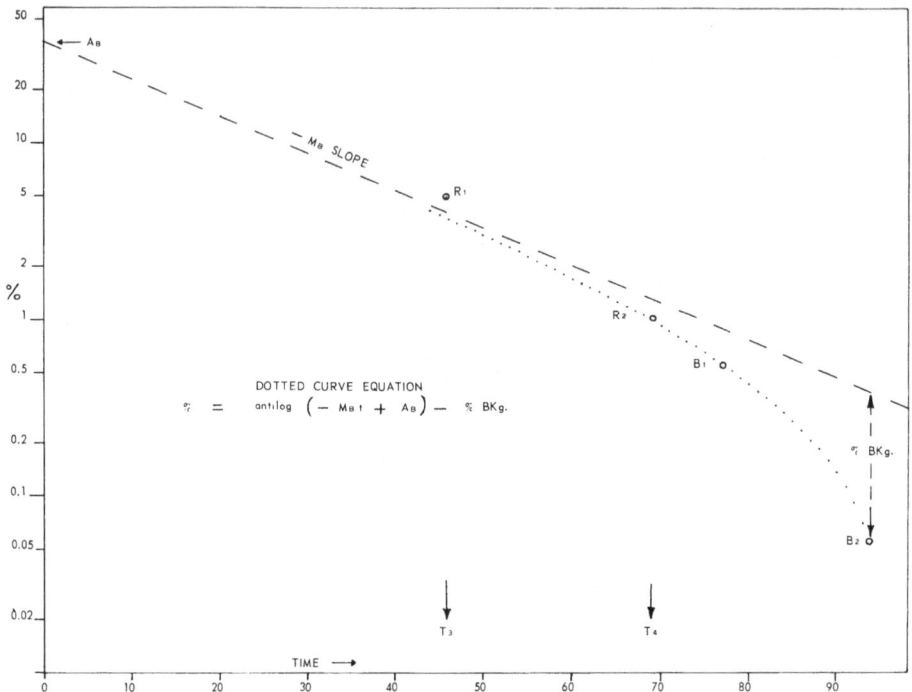

Figure 4. Background corrected portion of curve.

is no reversal, this is the final equation. However, notice points $R_1$ and $R_2$ in Fig. 4; they fall above the background corrected curve. They are standard samples which were run at the same time as points $B_1$ and $B_2$.

How can we correct our curve for these? If we look back at our capacitor voltages, as percent element goes up, it appears as though some voltage is leaking off the capacitor; since these points run too long, their voltages are too low. Or, the phototube current is becoming less (proportionately) as the element increases. Carrying this all the way back to the analytical gap, the line is being excited, but cool vapors surrounding the gap are absorbing the emission before the phototube sees it. So it appears just like a capacitor which has lost some of its charge, or that the element lost some of its percentage. Then it follows, that in the reversal end we must add something back to the background corrected curve to make the standard points agree. This something, not a constant this time, must increase with the number of atoms of that element. I instinctively felt that the correct added amount would be a power function, so Fig. 5 shows the reversal end of the curve (from actual samples) drawn above the computed

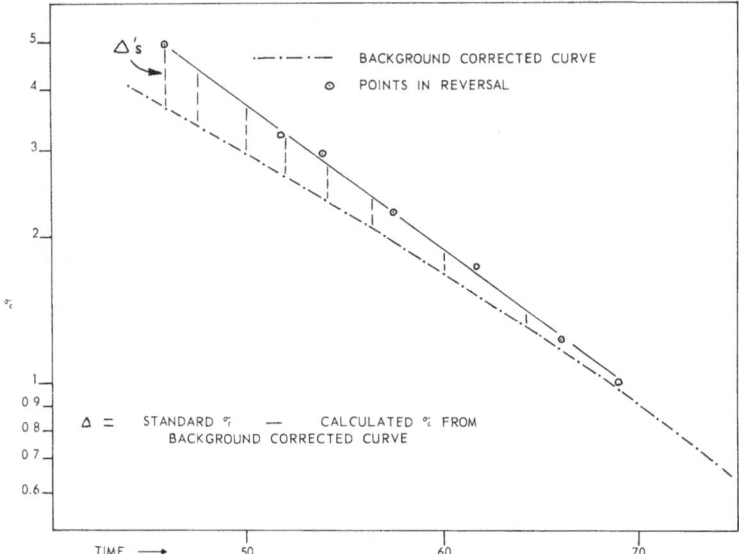

Figure 5. Reversal end of background corrected curve.

background corrected curve. The vertical lines joining the two curves, labeled deltas, represent the difference, in percent, between the two curves at each background corrected percent. I felt that the percent on the background corrected curve had to be raised to some power to equal the delta value to add to make the correction. From this, and other similar curves, I tested this theory as in Fig. 6. Here you see the background corrected percent versus the power to which it must be raised in order to obtain the correct delta to add. Notice the linear relationship. In attempting to use this relationship in computer form, everything worked fine, until a curve appeared where the background corrected percent became one percent, whereupon the power exponent made an abrupt change in sign, so the expression had to be made in another way. We have: ($\%$ = background corrected percent)

$$\Delta = (\%)^{M_R} \tag{2}$$

in rewriting the expression by taking the log of both sides we obtain

$$\log \Delta = M_R \log (\%) \tag{3}$$

Replotting the relationship on log–log paper, Fig. 7 typifies the result. Notice we again have a straight line, and this has been true with all cases where reversal appeared. Having established a working relationship in this

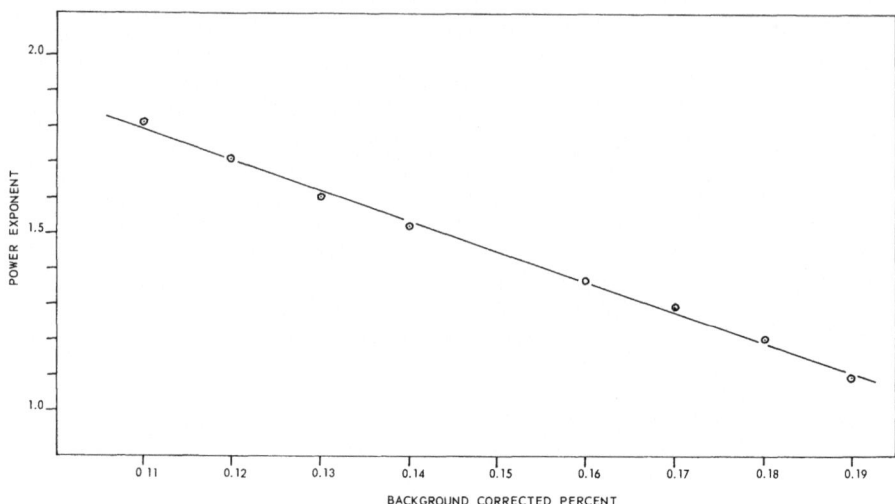

Figure 6.  Background corrected % *vs.* the power exponent.

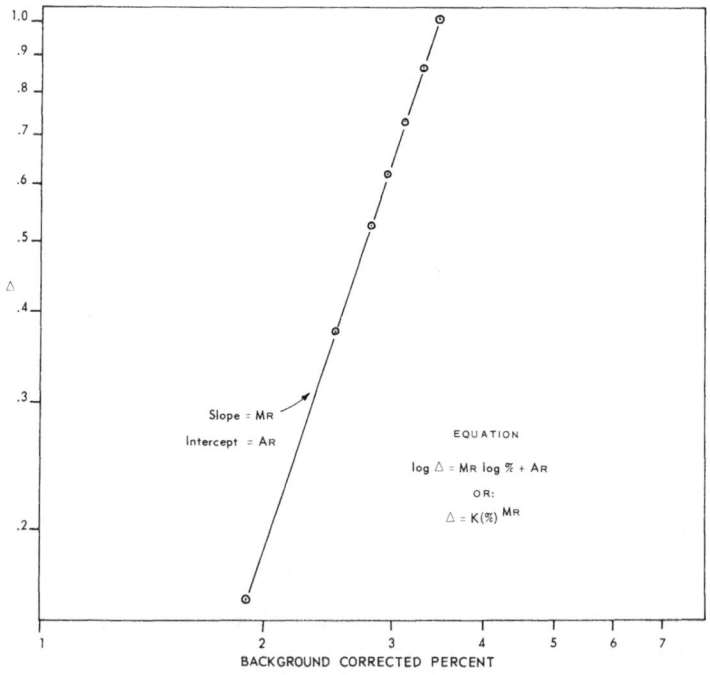

Figure 7.  Log background percent *vs.* log delta.

area, the computer program was expanded using two points in reversal to establish the delta values. Using these, the program calculates the slope and intercept of the log–log plot of background corrected percent *vs.* delta. I called the slope of this curve $M_R$ and the intercept was labeled $A_R$. The slope $M_R$ has units of log delta/log percent and $A_R$ is the log of the percent element being added at 1 % (background corrected percent).

This gives a complete description of the forces acting on the curve and leads to the equation for the curve as follows:

$$\% \text{ element} = (\text{antilog} [M_B(t) + A_B] - \text{BKG } \%)$$
$$+ \text{ antilog} [M_R(\log\{\text{antilog} [M_B(t) + A_B] - \text{BKG } \%\}) + A_R] \qquad (4)$$

Notice that the first term is the background corrected component of the curve and time is the only variable: the second term is the reversal component. An alternate form of the equation can be developed:

$$\text{Let: } X = 10^{M_B(t) + A_B} - \text{BKG } \% \qquad (5)$$

$$\text{Then: } \% = X^{M_R} 10^{A_R} + X \qquad (6)$$

Figure 8 shows a typical curve plotted from the equation using only the four points marked with a star to develop the computed constants. Circled points are additional standards, put in after the curve was constructed. The same standards appearing in Fig. 8 were submitted to regression fitting using all values, and the results show agreement, in general, to the nearest thousandth percent at every clock division with eighth degree regression.

As originally written, the computer program accepts the $x$ and $y$ values (time and standard percent) of four standards and generates the necessary constants. With several curves, it was apparent that the program was not accounting for the proportion of element being added (because of reversal) at point B-1. In other words, the plotted curve usually passed slightly high at this point. The magnitude of the deviation at this point (B-1) was used to further modify the program to plot the curve exactly through all four points. In applying the five constants in a real-time system to receive time as a variable and compute percent element, potential users will find that the constant $A_B$ will shift the curve parallel to itself without altering its shape. This constant is analogous to rotation of the clock face in Baird installations using clocks.

As with other methods using mathematics and computers, you must have good data to realize a good curve with four points. If the excitation is poor, or samples are bad and this data is given to the computer, it will give you a poor curve, or perhaps none at all, with the program as written.

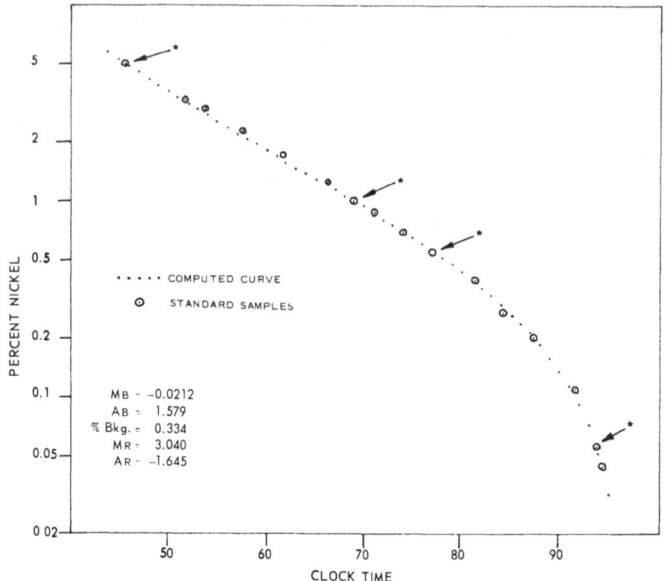

Figure 8. Spectrometer curve from equation.

The restrictions for the selected points are: two *must* be in background, and two *must* be in reversal. This may force people to use curves over broader concentration ranges than is generally the practice. It is *not* necessary to use only four points to define the curve. Actually, the curve may be obtained with polynominal regression and the four selected points taken from this curve and used to develop the five constants. Some have expressed concern at using only four points to obtain a curve. The main point of this paper is presentation of an equation for the spectrometer (or spectrographic) curve involving only five constants and only incidentally a way to develop the five constants from four selected standards.

In conclusion, the constants produced to identify the curve, in this paper, having meaning. An increase in reversal slope $M_R$ gives an increase in curvature at the reversal end of the curve. This steering constant, then, gives us a meaningful number to identify reversal and possibly classify excitation. The background percent to subtract gives a meaningful way to interpret the continuum under a line and a way to establish signal-to-background ratios, and limits of detection defined in terms of background variation. We intend to express our curves to the D... process computer in terms of these five constants.

Oh yes, before I forget, the title of this paper is: "Dear Computer, Where's My Curve?"

# REFERENCES

1. W. R. Kennedy, An Improvement to the Weekley–Norris Electronic Computer, *Appl. Spectry.* **21**, 338 (1967).

2. R. E. Weekley and J. A. Norris, A Versatile Electronic Computer for Photoelectric Spectrochemical Analysis, *Appl. Spectry.* **18**, 21 (1964).

3. W. R. Kennedy, Background Control in Direct Reading Analysis of Stainless Steels, *Appl. Spectry.* **19**, 74 (1965).

# INDEX

A/D converter, 19

Analysis of data at FORTRAN level, 3

ATR cell, 86

Assembly language program, 6

Background signal, 96

Baird capacitor discharge system, 94

Batch processing, 1, 11, 77

Coefficients for regression fit of instrumental data, 93

Computer monitoring of an infrared spectrophotometer, advantages of, 76–7

Computer–instrument systems
  off-line computer, gas chromatograph, 64
  on-line computer, gas chromatograph, 63
  on-line computer, infrared spectophotometer, 75
  composite integrator/computer, 64

Computers, by name
  Digital Equipment Corp., PDP-8, 40, 50
  IBM/1800, 2, 6, 8, 75
  IBM/360, 94, 97
  Infotronics CRS-110/50 Computer Integrator system, 63–73
  Olivetti 101, 94, 97
  Scientific Data Systems, Sigma 7, 19

Computers, by type
  built-in, 12
  dedicated, 1, 11, 13
  off-line, 12
  on-line, 11
  process control, 1
  time shared, 11, 13, 77

Control
  closed loop, 14, 82
  open loop, 13
  process, 77

Data
  acquisition, 77
    in real time, 2, 3, 18
    non-scheduled (demand) basis, 1
  analysis and reduction, 14, 77
  asynchronous encoding rate, 79
  filtering, 85
  at FORTRAN level, 2
  input rate, 2
  smoothing, 85

Detection of unresolved doublets, 31

Detectors, X-ray fluorescence, 43
  analyzing crystals, 43

Differential infrared spectra, 88

Digital output from computer, 8–10

Digital sweep system to control frequency or field of an NMR spectrometer, 49–62

Digitization of mass spectral data, on-line, 19

FORTRAN, 1–3, 5–6, 8–9, 77, 80,
   83, 91
Interfaces
  encoder, readout, Perkin-Elmer
    model 8RLS, 79
  goniometer control, 42
  instrument and computer, 3
Instrument
  automation, 1
  limitations, 5
  by name
    Associated Electrical Industries,
     Ltd., MS-902 high resolution
     mass spectrometer, 19
    Perkin-Elmer model 621 Infra-
     red Spectrophotometer, 75
    Siemens Kristalloflex IV con-
     stant potential generator and
     Universal sequential spec-
     trometer (SRS-1), 40
    Varian HR-60 NMR Spectrom-
     eter, 50
  by type
    exhaust emission, analyzer, 12
    gas chromatograph, 63–72
    real time, high resolution mass
     spectrometer, 18
Intensity, accuracy of measurement,
   25
Magnetic tape, chromatograms on,
   66–67
Mass measurement, accuracy of,
   21–22
Mass ratio, accuracy as a function of
   the number of scans, 29

Mass spectra, high resolution, 17
Multiplier gain as a function of ionic
   elemental composition, 21
Number of ions in a peak, 19
Phototube current, 98
Programs
  for analyzing gas chromatographic
    data, 69–71
  for controlling automatic X-ray
    fluorescence analysis, 44–46
Program maintenance, 4
Quality of a curve as the function of
   the number of data points, 101
Resolution of multiplets, 34–5
Signal-to-noise ratio, enhancement
   by averaging, 49
Smoothing function, criteria for
   selecting, 87
Spectrometer current, 95
Storage, of control programs, 51
Systems analyst, 2, 4
  design, 3
  documentation, 7
  maintenance, 4
System response time, 5–6
System software for time sharing, 92
Techniques
  infrared monitoring of a reaction
    system, 89
  X-ray fluorescence spectrography,
    39
Testing equipment, storage
   oscilloscope, 7
Time averaging of NMR signal, 50
Time Shared Executive (TSX), 8